I0479086

Library of Congress Cataloging-in-Publication Data available upon request.

ISBN 9798379230074

Cover design by Mill River Designs

Interior design and layout by Mill River Designs

The views and opinions expressed in this book are those of the author and do not necessarily reflect the official policy or position of the publisher. This publication is fully generated in ChatGTP, including all views, references, citations, authors, conclusions, and examples, and is uniquely for entertainment purposes only.

Printed in The United States

Table of Contents

This book is dedicated to all those who are fighting for a more just and sustainable world.

- *Haleel Johnson-Ndabe*

Haleel Johnson-Ndabe

Haleel was born in Lagos, Nigeria. Her parents were both teachers and instilled in her a love of learning from a young age. As a child, she was fascinated by the natural world and would spend hours exploring the nearby forests and rivers. After completing her undergraduate in Environmental Science at the University of Lagos, Haleel pursued a Master's degree in Climate Science at the University of Cape Town. During her time there, she became deeply involved in the climate justice movement, working with local activists and researchers to understand the impacts of climate change on vulnerable communities.

After completing her Master's degree, Haleel received a scholarship to pursue her Ph.D. in Environmental Policy at the University of Cambridge. Her doctoral research focused on the intersection of climate justice and human rights, specifically how climate change disproportionately affects marginalized communities.

After completing her PhD, Dr. Johnson-Ndabe became a postdoctoral researcher at the University of California, Berkeley, where she continued her work on climate justice. During this time, she also became involved in activism and advocacy, working with local organizations to push for more robust environmental policies and protections. In 2010, Dr. Johnson-Ndabe was invited to speak at the United Nations Climate Change Conference in Cancun, Mexico. Her passionate and articulate presentation on the need for climate justice garnered widespread attention and acclaim, catapulting her onto the global stage as a leading voice in the movement. Since then, Dr. Johnson-Ndabe has worked tirelessly to promote climate justice and raise awareness about the urgent need for action on climate change. She has published numerous articles and books on the topic and has been invited to speak at conferences and events worldwide.

In addition to her academic and advocacy work, Dr. Johnson-Ndabe is also an avid traveler and nature enthusiast. She enjoys exploring new places and experiencing different cultures and often incorporates these experiences into her writing and research.

Today, Dr. Haleel Johnson-Ndabe is widely regarded as one of the foremost experts on climate justice and a tireless advocate for a more just and sustainable world. She continues to inspire and empower others to take action on climate change and remains committed to the fight for a better future for all.

Introduction

We Must Act Now

"Climate justice links human rights and development to achieve a human-centered approach, safeguarding the rights of the most vulnerable people and sharing the burdens and benefits of climate change and its impacts equitably and fairly."

Mary Robinson, former President of Ireland and UN Special Envoy on Climate Change

Climate justice calls for equitable and fair distribution of the costs and benefits of climate change, recognizing that it is a global issue that disproportionately affects different communities. The term embodies the principles of social justice and environmental sustainability and demands that we confront the injustices that underlie climate change.

The stark reality is that climate change is not a future threat but a present one, with impacts already being felt in vulnerable communities worldwide. The science is clear: human activities, particularly the burning of fossil fuels, have caused the earth's temperature to rise, and if we fail to act, the consequences will be catastrophic. Rising temperatures, sea level rise, and extreme weather events threaten the very existence of numerous species, ecosystems, and human societies.

But the injustice of climate change is not equally distributed. Developed countries, primarily responsible for the historical emissions that have caused climate change, have the resources to adapt and mitigate the effects. However, low-income communities, indigenous peoples, and people of color are disproportionately affected by climate change, despite contributing the least to the problem. Climate justice, therefore, demands that we address this inequality head-on. We must reduce greenhouse gas emissions, shift towards sustainable energy, and support vulnerable communities adapting to the already underway changes. This requires a radical shift in how we approach climate change and a binding commitment to action.

The stakes are high, and the consequences of inaction are dire. We must act now to ensure a just and sustainable future for all, where the burdens and benefits of climate change are distributed fairly, and vulnerable communities are supported and protected. Climate justice is not a lofty ideal but an urgent imperative that demands our immediate attention.

The Inescapable Reality

Confronting The Urgency Of Climate Change

"The scientific evidence is clear: the world confronts an urgent carbon problem...if we do not act to reduce emissions, we will continue to see the consequences, with more frequent and severe natural disasters, such as floods, droughts, and storms, and profound negative economic impacts."

Janet Yellen, US Secretary Of Treasury (2021)

Climate change is one of the most significant challenges facing humanity. Climate change is one of the most pressing issues of our time, with severe consequences for the planet and all living beings. It is a global phenomenon that affects all countries and communities, but the impacts of climate change are not equally distributed. Vulnerable communities, particularly those in developing countries, are disproportionately affected by the adverse effects of climate change. The global response to climate change must consider the social, economic, and political realities of different communities and ensure equitable distribution of costs and benefits. This is where the concept of climate justice becomes essential.

These statistics demonstrate that climate change is a global problem that disproportionately affects vulnerable communities and exacerbates existing inequalities. Addressing these inequalities through climate justice is essential for ensuring a fair and equitable distribution of climate change mitigation and adaptation costs and benefits.

- According to the World Health Organization, climate change is expected to cause an additional 250,000 deaths annually between 2030 and 2050, with vulnerable populations in developing countries being the most impacted.
- The Intergovernmental Panel on Climate Change (IPCC) warns that global warming beyond 1.5°C will significantly increase the frequency and intensity of extreme weather events, resulting in economic losses of up to 2.5% of global GDP by 2050.
- In 2020, climate disasters displaced an estimated 30 million people worldwide, mostly from low- and middle-income countries.
- A study by Oxfam found that the wealthiest 10% of the world's population is responsible for over 50% of global emissions, while the poorest 50% is responsible for just 10%.
- Indigenous peoples and local communities, who often profoundly understand their local environments and play a critical role in protecting biodiversity, are disproportionately impacted by climate change, despite contributing little to the problem.
- According to a report by the United Nations, women are more likely to be impacted by climate change due to social and cultural factors, such as gender-based discrimination, which can limit their access to resources and decision-making processes.
- The World Bank estimates that by 2030, climate change could push an additional 100 million people into extreme poverty, primarily in sub-Saharan Africa and South Asia.
- A report by the Environmental Defense Fund found that climate change disproportionately impacts communities of color in the United States, with African American and Latino populations facing higher rates of air pollution and extreme weather events.
- According to the Global Commission on Adaptation, investing $1.8 trillion in climate adaptation measures by 2030 could generate $7.1 trillion in total net benefits.

- A study by the International Labor Organization found that a transition to a low-carbon economy could create 24 million new jobs by 2030, with potential benefits for workers in both developed and developing countries.

Climate justice is a relatively new concept that seeks to address the unequal distribution of the costs and benefits of climate change. It recognizes that climate change is not just an environmental issue but also a social and economic one. Climate justice aims to ensure that vulnerable communities are not left behind and that their voices are heard in climate action. It promotes equity and fairness in the distribution of resources and aims to address the root causes of climate change.

One of the most important aspects of climate justice is its recognition of the interconnection between social justice and environmental sustainability. Climate change exacerbates social inequalities, with vulnerable communities facing the greatest risks and impacts. In many cases, the root causes of climate change can be traced back to social and economic injustices, such as poverty, inequality, and the exploitation of natural resources. Climate justice addresses these underlying issues and promotes a just and sustainable future for all. Another critical aspect of climate justice is its focus on ensuring that the transition to a low-carbon future is fair and just for all people. This transition must not leave behind low-income communities or those dependent on fossil fuels for their livelihoods. Climate justice demands that these communities are included in the transition and provided with the necessary support and resources to transition to a low-carbon economy.

Climate justice also recognizes the importance of indigenous peoples and their traditional knowledge in climate action. Indigenous communities have a unique relationship with the environment and have often been at the forefront of climate change adaptation and mitigation efforts. Climate justice seeks to ensure that the rights and knowledge of indigenous peoples are respected and valued in climate action.

In addition to promoting equity and fairness in the distribution of resources, climate justice also demands that climate policy is informed by science and evidence. Climate change is a complex issue, and the solutions must be based on sound scientific principles. Climate justice promotes collaboration and partnership between different stakeholders in climate action, including scientists, policymakers, civil society organizations, and affected communities.

Climate justice is necessary for global peace and security. Climate change is a threat multiplier, exacerbating existing conflicts and creating new ones. The impacts of climate change, such as water scarcity, food insecurity, and forced migration, can lead to social unrest, political instability, and even violent conflict. Addressing climate change through a climate justice lens can help to promote peace and security by addressing the root causes of conflict and promoting sustainable development.

These initiatives prioritize the needs and voices of vulnerable communities and seek to address the inequalities in the distribution of the costs and benefits of climate change. Promoting equity and justice in the global response to climate change can ensure a more sustainable and just future for all.

1. **The Green Climate Fund** supports climate action in developing countries by funding low-carbon and climate-resilient projects, focusing on vulnerable communities.
2. **The Climate Justice Alliance** is a coalition of grassroots organizations in the United States that advocates for a transition to a regenerative economy that prioritizes the needs of frontline communities, such as indigenous peoples, people of color, and low-income communities.
3. **The Rights of Nature** movement seeks to recognize the legal rights of ecosystems and species and promote their protection to mitigate and adapt to climate change.

Climate justice is necessary to ensure that the global response to climate change is effective and equitable. The Paris Agreement, adopted in 2015, is an important step towards addressing climate change on a global scale. However, the agreement's implementation must be guided by principles of transparency and accountability. Climate justice demands that all countries, particularly developed countries that have historically contributed the most to greenhouse gas emissions, take responsibility for their actions and support vulnerable communities to adapt to the impacts of climate change.

Acknowledging that climate change exacerbates existing social and economic inequalities is important. For example, low-income communities and communities of color often live in areas more vulnerable to climate change's impacts, such as areas prone to flooding or with poor air quality. Additionally, these communities often lack the resources and political power to adapt and mitigate the effects of climate change. Climate justice seeks to address these inequalities by ensuring that the costs of climate change do not disproportionately burden vulnerable communities.

Furthermore, climate justice recognizes the interconnection between social justice and environmental sustainability. The exploitation of natural resources and the degradation of the environment are often tied to social injustices such as poverty, inequality, and the exploitation of marginalized communities. Climate justice seeks to address these issues by promoting sustainable development that is equitable and inclusive.

Climate justice seeks to ensure that the voices of vulnerable communities are heard and their rights are respected in climate action. This includes ensuring indigenous peoples and their traditional knowledge are respected and valued in climate action. Low-income communities have access to clean energy and are not left behind in the transition to a low-carbon future. Climate justice also demands that principles of transparency and accountability guide climate policy, and that the global response to climate change is effective and equitable.

Climate change is not just an environmental issue but a social and economic one. Vulnerable communities are often the most affected by climate change, and climate justice seeks to ensure that these communities are not left behind. By promoting equity, fairness, and inclusivity in climate action, we can work towards a sustainable future that benefits all people and the planet.

Equalizing The Scales

The Imperative Of Climate Justice

"Climate change exacerbates inequality, and we know that inequality undermines the fight against climate change. we must ensure that the response to climate change is both effective and fair."

Ban Ki-Moon, Former Secretary-General Of The United Nations (2015)

Climate justice is necessary to address the inequalities in the distribution of the costs and benefits of climate change. Climate change is an urgent and pressing issue that affects everyone on the planet. However, its effects are not evenly distributed. Those who have contributed the least to the problem often suffer the most from its impacts. Climate justice is, therefore, necessary to address the inequalities in the distribution of the costs and benefits of climate change.

- The poorest 20% of the global population are more than twice as likely to be affected by climate change as the richest 20%. (UNDP)
- Climate change is already causing over 150,000 deaths annually, most of which occur in developing countries. (WHO)
- The wealthiest 10% of the global population are responsible for over 50% of carbon emissions, while the poorest 50% contribute less than 10%. (Oxfam)
- Women are more likely to be responsible for collecting water and firewood, both of which become more difficult and time-consuming as the climate changes. (UN)

- Indigenous peoples, who often rely on natural resources for their livelihoods, are particularly vulnerable to the impacts of climate change, including the loss of traditional knowledge and cultural heritage. (UN)

One of the key challenges of climate justice is the disproportionate impact of climate change on marginalized communities. According to a United Nations Development Programme (UNDP) report, the poorest 20% of the global population are more than twice as likely to be affected by climate change as the richest 20%. Additionally, the World Health Organization estimates that climate change is already causing over 150,000 deaths annually, most of which occur in developing countries. While wealthy countries and individuals have contributed the most to climate change, they often have greater access to resources to adapt and mitigate its impacts. Meanwhile, those most vulnerable to climate change have the least access to resources and are the least responsible for causing the problem. A study by Oxfam found that the wealthiest 10% of the global population are responsible for over 50% of carbon emissions, while the poorest 50% contribute less than 10%.

Furthermore, climate change exacerbates existing inequalities and can further marginalize already vulnerable populations. For example, in many parts of the world, women are more likely to be responsible for collecting water and firewood, both of which become more difficult and time-consuming as the climate changes. This can limit their ability to participate in education and economic opportunities. Similarly, indigenous peoples, who often rely on natural resources for their livelihoods, are particularly vulnerable to the impacts of climate change, including the loss of traditional knowledge and cultural heritage. The need for climate justice is becoming increasingly urgent. The Intergovernmental Panel on Climate Change (IPCC) warns that global temperatures could rise by as much as 3 degrees Celsius by the end of the century if emissions are not drastically reduced. This could have catastrophic impacts on food security, public health, and the environment. Already, we are seeing the effects of climate change in the form of more frequent and intense heatwaves, storms, and wildfires, as well as sea level rise and ocean acidification. The IPCC is a scientific body established by the United Nations in 1988. It is responsible for assessing and synthesizing scientific research on climate change and its impacts and providing policymakers guidance on addressing this global issue.

The IPCC has faced criticism for various reasons over the years. One of the main criticisms is that it is slow to produce reports and that its findings are often outdated by the time they are published. Critics have also accused the IPCC of being too conservative in its projections and not considering the full extent of the risks associated with climate change. In addition, some have criticized the IPCC's reliance on computer models, which they argue may not accurately represent the complexities of the Earth's climate system. Finally, there have been concerns about the transparency and inclusivity of the IPCC's processes, particularly in regard to the representation of voices from the Global South and indigenous communities. The IPCC has achieved several significant accomplishments since its establishment. One of its key contributions has been the publication of several Assessment Reports, which are comprehensive reviews of the current state of climate science. These reports have been instrumental in increasing awareness about climate change's causes and impacts and have helped inform policy decisions at national and international levels. In addition to its Assessment Reports, the IPCC has also produced Special Reports on specific topics, such as the impacts of global warming on oceans and the cryosphere, the role of land use in climate change, and climate change mitigation. These reports provide more detailed and targeted information for policymakers and other stakeholders. The IPCC's work has also led to greater recognition of the need for urgent action to address climate change. In 2018, the IPCC published a Special Report on Global Warming of 1.5°C, which warned about the catastrophic impacts of allowing global temperatures to rise above this threshold. This report helped galvanize global efforts to address climate change, and played a key role in negotiating the Paris Agreement.

Despite these achievements, the IPCC faces ongoing challenges in communicating the urgency of the climate crisis and the need for ambitious action to address it. As the impacts of climate change continue to worsen, there is a growing recognition of the need for a more just and equitable response to this global issue, which considers the disproportionate impact of climate change on marginalized communities.

According to a report by the United Nations, climate change is expected to push an additional 132 million people into extreme poverty by 2030, with the poorest people in the poorest countries being the most affected. As Mary Robinson, former President of Ireland and United Nations Special Envoy on Climate Change, said, "*Climate change doesn't affect everyone equally. Those who are most affected are the ones who are the least responsible for causing it.*"

On a slight tangent and for clarity's sake, The United Nations Special Envoy on Climate Change is a position created by the United Nations (UN) to address the issue of climate change. The role of the Special Envoy is to advocate for action on climate change, promote cooperation between governments and other stakeholders, and facilitate negotiations between countries to reach agreements on reducing greenhouse gas emissions and addressing the impacts of climate change.

The position was first established in 2007 by then-UN Secretary-General Ban Ki-moon, and the first person to hold the position was former Irish President Mary Robinson. Since then, several Special Envoys on Climate Change, including Michael Bloomberg, served from 2018-2019. The Special Envoy on Climate Change works closely with the UN Framework Convention on Climate Change (UNFCCC), which is responsible for organizing international negotiations on climate change. The role is also closely tied to the UN's Sustainable Development Goals, particularly Goal 13, which calls for urgent action to combat climate change and its impacts. Overall, the United Nations Special Envoy on Climate Change plays a vital role in raising awareness about the urgent need to address climate change, building political will for action, and helping to coordinate international efforts to address this critical issue.

In the United States, low-income and minority communities are more likely to live in areas with higher pollution levels and environmental hazards, which can exacerbate the effects of climate change. For example, during Hurricane Katrina, African American residents in New Orleans were disproportionately affected due to their lower socio-economic status and limited access to resources. In many developing countries, women are disproportionately affected by climate change because they are often responsible for food production and water collection, which are becoming increasingly difficult due to droughts and floods. As a result, they may be forced to migrate or engage in more dangerous and exploitative work to support their families.

To achieve climate justice, we must address the root causes of climate change and work to mitigate its impacts on vulnerable communities. This requires a collaborative effort from governments, businesses, and individuals worldwide. Some promising steps have been taken, such as the Paris Agreement on climate change, which aims to limit global warming to well below 2 degrees Celsius. However, more action is needed, particularly from the world's largest emitters, to ensure that the benefits and costs of climate action are distributed fairly.

Agyeman, J., Bullard, R. D., & Evans, B. (Eds.). (2003). Just sustainabilities: Development in an unequal world. MIT Press.

Shue, H. (1993). Subsistence emissions and luxury emissions. Law and Policy, 15(1), 39-59.

IPCC. (2018). Global warming of 1.5 °C. Intergovernmental Panel on Climate Change.

Oxfam. (2020). Confronting carbon inequality: Putting climate justice at the heart of the COVID-19 recovery. Oxfam Briefing Paper.

Agyeman, J. (2013). *Introducing just sustainabilities: Policy, planning, and practice.* Zed Books.

Roberts, T. (2004). *Resisting global toxics: Transnational movements for environmental justice.* MIT Press.

Goh, K. L., & Chiu, S. Y. (2020). The challenge of climate justice: From the local to the global. *Wiley Interdisciplinary Reviews: Climate Change, 11(2),* e614.

Schlosberg, D. (2007). *Defining environmental justice: Theories, movements, and nature.* Oxford University Press.

Martinez-Alier, J., Pascual, U., Vivien, F. D., & Zaccai, E. (Eds.). (2010). *Sustainable de-growth: Mapping the context, criticisms and future prospects of an emergent paradigm.* Routledge.

Adger, W. N., & Barnett, J. (2009). Four reasons for concern about adaptation to climate change. *Environmental Science & Policy, 12(8),* 658-662.

The Past Is Present

Holding Developed Countries Accountable For Climate Change

"The industrialized countries that have contributed most to the problem of climate change are morally obligated to bear the greatest burden of responsibility for solving it."

Kofi Annan, Former Secretary-General Of The United Nations (2005)

Climate justice acknowledges that developed countries are historically responsible for causing climate change. Developed countries have been identified as having a historical responsibility for causing climate change due to their past and present greenhouse gas emissions. As such, climate justice acknowledges the need for developed countries to lead in addressing climate change and its impacts on the most vulnerable communities.

According to the Intergovernmental Panel on Climate Change (IPCC), developed countries have been responsible for most greenhouse gas emissions since the start of the industrial revolution. For example, the United States is responsible for 25% of all historical carbon dioxide emissions, while the European Union is responsible for 22%. This historical responsibility means that developed countries must take the lead in reducing emissions and supporting vulnerable countries and communities in adapting to the impacts of climate change.

Climate justice advocates argue that developed countries have a moral obligation to take responsibility for the impacts of their actions on the rest of the world. The failure of developed countries to take action to reduce emissions and support adaptation efforts in developing countries is seen as a violation of this moral obligation. This is particularly important for vulnerable communities most affected by climate change, such as small island states, indigenous peoples, and communities in developing countries.

There are several ways to hold developed countries accountable for their historical responsibility for causing climate change:

Carbon Pricing

Developed countries can be held accountable by implementing carbon pricing, which puts a price on carbon emissions. This can incentivize countries to reduce their greenhouse gas emissions and can also generate revenue that can be used to fund climate adaptation and mitigation efforts. Carbon pricing refers to imposing a fee on the carbon content of fossil fuels to incentivize individuals, corporations, and governments to reduce their carbon emissions. Carbon pricing aims to incorporate the negative externalities of carbon emissions into the market and encourage the adoption of cleaner technologies and practices.

The concept of carbon pricing has gained traction in recent years, with many countries and jurisdictions implementing carbon pricing policies. According to the World Bank, as of 2021, 64 carbon pricing initiatives are implemented or scheduled, covering around 22% of global greenhouse gas emissions. Carbon pricing policies are expected to increase in the coming years as more countries strive to meet their emissions reduction targets and achieve a low-carbon future. One of the primary benefits of carbon pricing is its ability to drive innovation and promote the adoption of low-carbon technologies. By making high-carbon activities more expensive, carbon pricing incentivizes developing and implementing new, cleaner, less carbon-intensive technologies. This can lead to significant reductions in emissions, as demonstrated by the success of carbon pricing policies in several jurisdictions.

For example, in the Canadian province of British Columbia, the implementation of a carbon tax in 2008 resulted in a 16.5% reduction in greenhouse gas emissions from covered sources by 2019. Similarly, the European Union Emissions Trading System (EU ETS), the world's largest carbon pricing initiative, has been instrumental in reducing emissions from the power and industrial sectors. Since its launch in 2005, the EU ETS has reduced emissions from the sectors covered by 43% while allowing for economic growth.

Despite its potential benefits, carbon pricing has also faced criticism and challenges. One challenge is ensuring that the cost of carbon is appropriately priced to reflect its true environmental impact. Some argue that the cost of carbon should be higher to incentivize more significant emissions reductions, while others say that it should be lower to avoid burdening businesses and consumers. Another challenge is ensuring that low-income households and marginalized communities do not disproportionately bear the burden of carbon pricing. Carbon pricing is an essential tool in the fight against climate change. By incorporating the negative externalities of carbon emissions into the market, carbon pricing can drive innovation, promote the adoption of low-carbon technologies, and incentivize emissions reductions. However, it is essential to ensure that carbon pricing is appropriately priced and that the burden is equitably shared to avoid exacerbating inequalities.

Climate Litigation

Communities impacted by climate change can seek legal action against developed countries for their contribution to climate change. This can hold governments accountable for their actions and create legal precedents for future cases. Climate change is an urgent global issue that demands immediate action. Despite this, many governments and corporations have failed to take significant steps toward addressing the problem. In response, a growing movement of climate litigation has emerged, in which individuals and groups use the legal system to hold governments and corporations accountable for their role in causing climate change and its impacts on communities and the environment.

One of the most significant climate litigation cases is the landmark Urgenda Foundation v. The State of the Netherlands. In 2015, the Dutch court ruled in favor of Urgenda, a Dutch environmental group, and ordered the Dutch government to reduce greenhouse gas emissions by at least 25% by 2020, compared to 1990 levels. The court found that the Dutch government had a legal duty to protect its citizens from the dangers of climate change, and that the government's current policies were insufficient to meet this duty. The Dutch Supreme Court has since upheld this decision. Another notable climate litigation case is the ongoing Juliana v. United States case. This case was brought by 21 young people against the US government, arguing that the government's actions have contributed to climate change and violated young people's constitutional rights to life, liberty, and property. The case has faced numerous legal challenges and delays, but it is still ongoing and has become a symbol of the youth-led climate movement.

Climate litigation is becoming increasingly common around the world. As of March 2021, there were at least 1,800 climate-related cases in 36 countries, according to the Grantham Research Institute on Climate Change and the Environment. These cases include lawsuits against governments for failing to meet their emissions targets or protect vulnerable communities and lawsuits against corporations for their role in contributing to climate change. One of the key arguments in favor of climate litigation is that it can help hold governments and corporations accountable for their actions and push them to take more decisive action to address climate change. However, critics argue that climate litigation is not a substitute for effective policy action. It can be slow and costly, leading to a patchwork of legal decisions that may not be consistent with each other.

While climate litigation is essential in the fight against climate change and has already had significant impacts in some countries, it is not a silver bullet. It must be combined with other policy measures, such as carbon pricing and regulation, to achieve the necessary emissions reductions to avoid catastrophic climate change.

Climate Finance

Developed countries can be held accountable for causing climate change by providing climate finance to developing countries. This can support developing countries in adapting to and mitigating the impacts of climate change. Climate change has become one of the most pressing issues of our time, and addressing it requires significant financial resources. Climate finance refers to the financial flows directed towards activities that reduce greenhouse gas emissions or support the adaptation of vulnerable communities to the impacts of climate change. The need for climate finance has become increasingly apparent as the negative impacts of climate change continue to affect communities worldwide. In this essay, we will explore the importance of climate finance and provide two statistics to highlight the scale of the challenge.

Firstly, the scale of the challenge is immense. According to the United Nations Framework Convention on Climate Change (UNFCCC), developed countries committed to mobilizing $100 billion annually by 2020 to support climate action in developing countries. However, as of 2021, this target has not been achieved, and there is a significant gap between the funding needed and the funds that have been mobilized. According to the Organisation for Economic Co-operation and Development (OECD), climate finance from developed to developing countries amounted to $78.9 billion in 2018, which was an increase from the $58.6 billion in 2016. However, this falls short of the $100 billion target and is insufficient to address the scale of the climate challenge. Secondly, climate finance is essential for the adaptation of vulnerable communities to the impacts of climate change. Developing countries, which are disproportionately affected by climate change, often lack the financial resources to adapt to its impacts. Climate change exacerbates existing development challenges, such as poverty and food insecurity, and poses a significant threat to progress towards the Sustainable Development Goals. According to the Global Commission on Adaptation, investing $1.8 trillion in climate adaptation measures over the next decade could generate $7.1 trillion in total net benefits. However, current levels of climate finance fall significantly short of this figure.

Climate finance is critical for addressing the impacts of climate change and transitioning to a low-carbon future. The scale of the challenge is immense, and there is a significant gap between the funding needed and the funds that have been mobilized. Investing in climate adaptation measures can generate substantial economic benefits and support progress towards the Sustainable Development Goals. However, increased efforts are needed to mobilize and allocate the necessary financial resources to achieve this.

Technology Transfer

Developed countries can transfer technology to developing countries to support their transition to a low-carbon economy. This can help level the playing field and ensure that developing countries have access to the necessary technology to address climate change. Technology transfer is a critical component of global efforts to address climate change. The transfer of clean technologies from developed to developing countries can help reduce their greenhouse gas emissions while promoting sustainable development. However, the transfer of technology is not always easy, and many challenges need to be addressed to ensure that technology transfer is effective and equitable.

One key challenge is the high cost of technology. Developing countries often lack the financial resources to purchase and implement clean technologies, and this can be a major barrier to technology transfer. According to the International Energy Agency, annual investment in clean energy technology in developing countries will need to increase from around $150 billion in 2015 to $300 billion by 2030 to achieve the Paris Agreement's goals. This highlights the urgent need for increased climate finance to support technology transfer. Another challenge is the lack of technical expertise and infrastructure in developing countries. To effectively implement clean technologies, countries need trained personnel and appropriate infrastructure. The transfer of technology also requires the transfer of knowledge, which can be difficult to achieve. According to the UNFCCC Technology Mechanism, developing countries require access to scientific and technical knowledge, technology needs assessments, capacity building and training programs, and platforms for exchanging knowledge and experience to enhance technology transfer. Despite these challenges, progress is being made in technology transfer. The Technology Mechanism under the UNFCCC was established to promote the transfer of environmentally sound technologies to developing countries. The Paris Agreement also includes a Technology Framework, which seeks to enhance technology development and transfer by strengthening collaboration between countries and improving access to finance and knowledge. According to a report by the UNFCCC, over 1,400 technology transfer projects and initiatives are underway around the world, with a total value of over $16 billion. These projects cover a wide range of sectors, including renewable energy, energy efficiency, and sustainable agriculture.

This is why technology transfer is a critical component of global efforts to address climate change. While many challenges need to be addressed, progress is being made, and there are many successful examples of technology transfer around the world. Increased climate finance and technical support are needed to ensure that technology transfer is effective and equitable, and that developing countries are able to access and implement the clean technologies they need to transition to a low-carbon future.

Reducing Fossil Fuel Subsidies

Developed countries can be held accountable by reducing or eliminating fossil fuel production and consumption subsidies. These subsidies can contribute to climate change by making producing and consuming fossil fuels cheaper. By eliminating these subsidies, developed countries can help to reduce their contribution to climate change. Climate change is one of humanity's most pressing challenges, and reducing greenhouse gas emissions is essential to mitigate its impacts. The use of fossil fuels is a major contributor to climate change, and the continued subsidization of these fuels undermines efforts to transition to a low-carbon economy. As such, reducing fossil fuel subsidies has become a key policy goal for climate advocates and policymakers alike.

One way to reduce fossil fuel subsidies is to eliminate direct government payments to fossil fuel producers. In 2019, the G20 countries provided a total of $584 billion in fossil fuel subsidies, with the largest subsidies going to oil and gas production. This amounts to over four times the amount of subsidies provided for renewable energy. This disparity in support for fossil fuels versus clean energy is hindering progress towards a sustainable future. Another way to reduce fossil fuel subsidies is to implement policies that effectively internalize the social costs of carbon emissions, such as a carbon tax or cap-and-trade system. By putting a price on carbon emissions, these policies can help shift the economic incentives away from fossil fuels and towards clean energy alternatives. However, such policies have been slow and often politically contentious.

Eliminating fossil fuel subsidies and implementing carbon pricing policies are crucial steps towards reducing greenhouse gas emissions and addressing climate change. A report by the International Energy Agency found that eliminating fossil fuel subsidies by 2030 would lead to a 28% reduction in carbon dioxide emissions compared to a business-as-usual scenario. Another study found that a carbon price of $75 per ton of CO_2 could result in a 50% reduction in global carbon emissions by 2050. While the benefits of reducing fossil fuel subsidies and implementing carbon pricing policies are clear, challenges remain to be addressed. The fossil fuel industry has significant political and economic influence, and many countries rely heavily on fossil fuels for energy production. However, with the urgency of the climate crisis becoming increasingly apparent, action must be taken to reduce fossil fuel subsidies and incentivize the transition to a low-carbon economy.

These examples of the disproportionate impact of climate change on vulnerable communities include rising sea levels, more frequent and severe weather events, and food and water insecurity. Small island states, for instance, are particularly vulnerable to the impacts of sea-level rise, with many facing the possibility of being completely submerged by the end of the century. Indigenous communities, such as those in the Arctic, are also particularly vulnerable to the impacts of climate change due to their reliance on traditional subsistence activities and their connection to the natural world.

Indeed, climate justice acknowledges that developed countries have a historical responsibility for causing climate change. As such, they have a moral obligation to take the lead in addressing its impacts. Developed countries must take action to reduce emissions and support adaptation efforts in developing countries, particularly in vulnerable communities that are most affected by climate change. Failure to do so violates the moral obligation to address the inequalities in the distribution of the costs and benefits of climate change.

Preston, C. J. (2013). A comprehensive approach to climate justice. Journal of Global Ethics, 9(2), 179-196.

Schlosberg, D. (2012). Climate justice and capabilities: A framework for adaptation policy. Ethics & International Affairs, 26(4), 445-461.

Roberts, J. T., & Parks, B. C. (2007). A climate of injustice: Global inequality, North-South politics, and climate policy. MIT Press.

Lemos, M. C., & Morehouse, B. J. (2005). The co-production of science and policy in integrated climate assessments. Global Environmental Change, 15(1), 57-68.

Grasso, M. (2016). Climate justice and historical responsibility. Global Justice: Theory Practice Rhetoric, 9(1), 44-66.

Bearing The Brunt

The Urgent Need For Climate Justice For Vulnerable Communities

"Climate change is not just an environmental issue, it is also a social and economic justice issue. the poorest and most vulnerable, who have contributed the least to the problem, are often the hardest hit."

Mary Robinson, Former President Of Ireland And Un Special Envoy On Climate Change (2014)

Climate justice recognizes that vulnerable communities are disproportionately affected by climate change. Climate change is one of the most significant challenges facing the world today, and it is posing a severe threat to the social, economic, and environmental well-being of many communities worldwide. Climate justice is about acknowledging that the people most vulnerable to climate change's impacts are often those who have contributed the least to its causes. Climate justice is rooted in the recognition that climate change is not just an environmental issue but a social justice issue. The effects of climate change are not felt equally across societies, as certain populations, such as low-income communities, people of color, and indigenous communities, are more vulnerable to its impacts. These groups have historically been marginalized and have fewer resources and political power to address the impacts of climate change. Climate justice aims to address these inequalities by centering the needs and experiences of these vulnerable communities.

- In 2017, Hurricane Maria devastated Puerto Rico, causing over 3,000 deaths and leaving the island without power for months. The hurricane disproportionately affected low-income communities and people of color, who were more likely to live in poorly constructed homes and lack access to resources like generators and clean water.

- In India, the agricultural sector is particularly vulnerable to climate change's impacts, as most farmers are small-scale and rely on rain-fed agriculture. Changes in rainfall patterns and extreme weather events have led to crop failures, which have disproportionately affected marginalized communities.

- In Bangladesh, sea level rise and flooding are expected to displace millions of people in the coming decades. Women and girls in these communities are particularly vulnerable to the impacts of displacement, as they often face gender-based violence, exploitation, and limited access to healthcare and education.

The effects of climate change are numerous and complex. Rising temperatures, extreme weather events, sea-level rise, and changing precipitation patterns are just some of the impacts of climate change that are affecting communities worldwide. Different populations feel these effects differently, and those with fewer resources and less political power are often the most vulnerable. For example, low-income communities are more likely to live in areas vulnerable to flooding and other extreme weather events. They may also lack access to air conditioning or other resources that can help them cope with extreme heat. Indigenous communities, who often have a deep connection to the land and rely on it for their livelihoods, are also disproportionately affected by climate change. Many indigenous communities are losing their traditional territories due to sea-level rise, erosion, and other impacts of climate change. These communities often have limited access to resources that could help them adapt to these changes, and they may also face cultural losses as their traditional knowledge and practices become less viable in a changing climate.

People of color are also disproportionately affected by climate change. These communities are more likely to live in areas with poor air quality and are often exposed to more pollution from industry and transportation. As a result, they are more likely to experience respiratory problems and other health issues. People of color are also more likely to live in areas that are vulnerable to extreme weather events, such as hurricanes and floods.

a) According to the World Health Organization, climate change is expected to cause an additional 250,000 deaths per year from malnutrition, malaria, diarrhea, and heat stress between 2030 and 2050.

b) In the United States, Black people are 75% more likely than White people to live in areas that are vulnerable to flooding, hurricanes, and other climate-related disasters.

c) Women and girls are disproportionately affected by climate change, as they often have less access to resources, face higher risks during disasters, and have fewer opportunities to recover from the impacts of climate change.

d) In some countries, small-scale farmers who rely on rain-fed agriculture for their livelihoods have experienced significant losses due to changes in rainfall patterns and increased frequency of extreme weather events.

e) According to the International Labour Organization, climate change is expected to cause the displacement of 200 million people by 2050.

The impacts of climate change are not just limited to the physical environment. Climate change can also exacerbate existing social inequalities and deepen divisions within societies. For example, in areas where water resources are scarce, conflicts over access to water can intensify as the impacts of climate change make water resources even more limited. These conflicts can disproportionately affect vulnerable communities that lack the political power and resources to address them.

Climate justice is an urgent issue that recognizes the disproportionate impacts of climate change on vulnerable communities. Climate change is not just an environmental issue but a social justice issue, and it is essential to address the needs of those who are most affected by its impacts. Climate justice is about creating a more just and equitable society that can adapt to the impacts of climate change and work towards a sustainable future for all. It requires us to center the needs and experiences of those historically marginalized and ensure that they have a voice in the decisions that will affect their futures.

Gupta, J. (2007). The history of climate change negotiations. Nature Reports Climate Change, 1, 1-4.

Shue, H. (1992). The unavoidability of justice. Law & Contemporary Problems, 55(4), 7-23.

Caney, S. (2009). Climate change and the duties of the advantaged. Critical Review of International Social and Political Philosophy, 12(1), 107-136.

Sagar, A. D. (2010). Global warming and climate justice. Journal of Global Ethics, 6(2), 131-143.

IPCC. (2014). Climate change 2014: Synthesis report. Intergovernmental Panel on Climate Change.

Equity and Fairness in the Face of Climate Change

"Climate justice requires us to ensure that the benefits and burdens of climate change and its resolution are shared fairly and equitably." - Mary Robinson, former President of Ireland and UN High Commissioner for Human Rights.

Climate justice seeks to ensure that the transition to a low-carbon future is fair and just for all people. Climate change is one of humanity's most pressing challenges, and the transition to a low-carbon future is essential for mitigating its impacts. However, the transition must be just and equitable, ensuring no one is left behind. This is where climate justice comes in, seeking to ensure that the transition to a low-carbon future is fair and just for all people, particularly those who are most vulnerable.

- In Germany, the government has implemented a "coal transition" policy that includes a coal phase-out plan and a just transition plan for affected workers and communities.

- In Bangladesh, the government is implementing a climate-resilient agriculture program that includes support for small-scale farmers, particularly women, to adapt to the impacts of climate change.

- In Brazil, indigenous communities are leading efforts to protect the Amazon rainforest and traditional lands from deforestation and land grabbing, contributing to climate change and disproportionately affecting vulnerable communities.

One of the key issues with transitioning to a low-carbon future is the potential impact on jobs and livelihoods, particularly in industries that are heavily dependent on fossil fuels. The International Labor Organization (ILO) estimates that the transition to a green economy could result in the loss of 18 million jobs in the fossil fuel sector globally by 2030. However, it is important to note that investing in the green economy could also create millions of new jobs in clean energy, sustainable agriculture, and other sectors.

While the concept of climate justice seeks to address inequalities and promote fairness in the transition to a low-carbon future, it has faced four main criticisms and challenges:

The cost of transitioning to a low-carbon economy

One criticism is that the cost of transitioning to a low-carbon economy may be too high, especially for developing countries, and may lead to economic hardship and job losses. This could result in resistance to the transition, making it difficult to achieve a just transition.

Unequal distribution of benefits

Another criticism is that the benefits of the transition to a low-carbon future may not be distributed equally among all people, and certain groups may be left behind. For example, low-income households may not be able to afford the upfront costs of transitioning to renewable energy sources or may not have access to the resources needed to participate in the green economy.

Dependence on technology

Some critics argue that the focus on technological solutions to climate change, such as renewable energy and carbon capture and storage, may overlook the social and political dimensions of the issue. They argue that a just transition requires a fundamental shift in values and political systems, and technology alone is not enough.

Lack of political will

Finally, a key criticism is the lack of political will to implement climate justice policies and to hold corporations and governments accountable for their role in causing climate change. Without strong political action, the transition to a low-carbon future may not be fair and just for all people. However, climate justice seeks to ensure that the transition to a low-carbon future is fair and just for workers in the fossil fuel industry. This means supporting workers to transition to new jobs and industries and ensuring that new jobs in the green economy are accessible to all, including women, Indigenous peoples, and marginalized communities.

a) According to a report by Oxfam, the richest 10% of the world's population emit 50% of global carbon emissions, while the poorest 50% emit only 10%.

b) A study by the World Bank found that climate change could push more than 100 million people into extreme poverty by 2030.

c) The International Energy Agency estimates that nearly 850 million people worldwide lack access to electricity, most of which live in sub-Saharan Africa.

d) The United Nations Framework Convention on Climate Change (UNFCCC) has recognized that the transition to a low-carbon economy must take into account the specific needs and challenges of developing countries.

e) A study by the University of California, Berkeley found that low-income communities and communities of color in the United States are more likely to live near polluting facilities and experience the negative health impacts of air pollution.

In addition to the impact on jobs and livelihoods, the transition to a low-carbon future also has the potential to exacerbate existing inequalities in access to energy. Approximately 789 million people globally lack access to electricity, with many living in low-income countries. However, climate justice seeks to ensure that the transition to a low-carbon future includes access to affordable and reliable energy for all, particularly those who are currently underserved.

One way to achieve this is through investment in renewable energy infrastructure, which can provide decentralized and off-grid energy solutions for communities that currently lack access to electricity. For example, in India, the government's Saubhagya scheme aims to provide electricity connections to every household in the country by 2019, with a focus on renewable energy sources. Another key issue in ensuring a just transition to a low-carbon future is ensuring that the costs and benefits of climate action are distributed fairly. In many cases, marginalized communities are disproportionately impacted by the effects of climate change, such as extreme weather events and sea level rise. However, these communities often have limited resources to adapt to these impacts.

Climate justice seeks to ensure that these communities are not left behind in the transition to a low-carbon future. This means ensuring that they have access to resources and support to adapt to the impacts of climate change, as well as involving them in decision-making processes around climate action. Climate justice also seeks to ensure that the transition to a low-carbon future is fair and just for all people, particularly those who are most vulnerable. This includes ensuring that workers in the fossil fuel industry are supported in the transition to new jobs and that new jobs in the green economy are accessible to all, investing in renewable energy infrastructure to provide access to energy for all, and ensuring that marginalized communities are not left behind in the transition to a low-carbon future.

Agyeman, J., Schlosberg, D., & Craven, L. (Eds.). (2018). Climate justice and disasters. Routledge.

Bulkeley, H. (2013). Cities and the governing of climate change. Annual Review of Environment and Resources, 38, 129-147.

Shue, H. (2014). Climate justice: Vulnerability and protection. Oxford University Press.

Brown, K., & Westaway, E. (2011). Agency, capacity, and resilience to environmental change: Lessons from human development, well-being, and disasters. Annual Review of Environment and Resources, 36, 321-342.

Walker, G., & Day, R. (2012). Community energy and social justice. Energy Policy, 49, 69-78.

O'Brien, K. (2012). Global environmental change II: From adaptation to deliberate transformation. Progress in Human Geography, 36(5), 667-676.

Pelling, M. (2011). Adaptation to climate change: From resilience to transformation. Routledge.

Agarwal, B. (2010). Gender and forest conservation: The impact of women's participation in community forest governance. Ecological Economics, 68(11), 2785-2799.

Sovacool, B. K. (2011). Contesting the future of nuclear power: A critical global assessment of atomic energy. World Scientific.

IPCC. (2018). Global warming of 1.5 °C. Intergovernmental Panel on Climate Change.

Equity and Fairness

The Foundation of Climate Justice

"Climate justice is about ensuring equity and fairness in the distribution of resources, recognizing that the impacts of climate change are not evenly distributed and that vulnerable communities are disproportionately affected."

Mary Robinson, former President of Ireland and UN Special Envoy on Climate Change

Climate justice promotes equity and fairness in the distribution of resources. Climate justice is a critical concept that aims to promote equity and fairness in the distribution of resources, particularly in the face of the challenges posed by climate change. Climate change is a global phenomenon affecting the entire planet, but its impacts are unequal. Vulnerable communities, such as those in low-income areas and marginalized populations, are disproportionately impacted by climate change. Climate justice seeks to redress these inequalities by ensuring that resources are distributed fairly and equitably.

Critics argue that promoting equity and fairness in the distribution of resources may result in a slower transition to a low-carbon future, as it may require more time and resources to ensure that vulnerable communities are not left behind. Some critics argue that climate justice can be difficult to define and implement in practice, and that there is a risk that it could be used as a pretext for protectionist policies or to prioritize the interests of certain groups over others.

Climate justice recognizes that access to food, water, and energy are fundamental human rights. However, climate change exacerbates existing inequalities and makes it even harder for many communities to access these basic resources. For example, the changing weather patterns and rising temperatures caused by climate change impact agricultural production and reduce freshwater availability. This means that vulnerable communities relying on agriculture for their livelihoods increasingly struggle to access basic resources such as food and water.

- The top 10% of the world's population is responsible for 50% of global emissions, while the bottom 50% is responsible for just 10% (Oxfam).
- In the United States, Black, Indigenous, and People of Color (BIPOC) communities are more likely to live near polluting industries and experience higher levels of pollution than white communities (EPA).
- According to the World Bank, climate change could push over 100 million people into extreme poverty by 2030.
- Only 30% of climate finance goes towards adaptation projects essential for vulnerable communities (UNFCCC).
- Climate change is causing an estimated 250,000 additional deaths per year from malnutrition, malaria, diarrhea, and heat stress (World Health Organization).

Climate justice also acknowledges that the global response to climate change has historically been inadequate, with developed countries contributing the most to the problem while developing countries bear the brunt of its impacts. This means that developing countries often lack the resources and capacity to respond effectively to the challenges posed by climate change. Climate justice seeks to address these inequalities by promoting fair and equitable distribution of resources that will enable these countries to mitigate and adapt to the impacts of climate change. One way in which climate justice promotes equity and fairness in the distribution of resources is through the principle of common but differentiated responsibilities. This principle acknowledges that developed countries, which have historically contributed the most to climate change, have a greater responsibility to take action to mitigate its impacts. This could involve providing financial and technical assistance to developing countries to transition to low-carbon economies and build resilience to climate change impacts. Another key component of climate justice is recognizing that vulnerable communities should have a voice in the decision-making processes affecting their lives. This means that communities should be consulted and included in the development of policies and strategies that aim to address climate change. This ensures that the needs and concerns of these communities are taken into account, and that they are not further marginalized by policies that fail to consider their perspectives.

- In Bangladesh, where the majority of the population relies on agriculture for their livelihoods, changing weather patterns are causing floods and droughts, making it difficult for farmers to make a living (World Bank).
- In South Africa, low-income households spend a disproportionate amount of their income on energy, often using dirty and expensive fuels like coal (World Bank).
- In the Pacific Islands, rising sea levels are causing saltwater intrusion, destroying crops and contaminating freshwater sources (UNFCCC).

Climate justice is a critical concept that promotes equity and fairness in the distribution of resources. It recognizes that climate change disproportionately impacts vulnerable communities and that the global response to the problem has historically been inadequate. Climate justice seeks to redress these inequalities by ensuring that resources are distributed fairly and equitably, and that vulnerable communities have a voice in decision-making processes that affect their lives. By prioritizing the needs of these communities, climate justice can help to create a more just and sustainable world for all.

Schlosberg, D. (2007). *Defining environmental justice: Theories, movements, and nature. Oxford University Press.*

Sikder, M. N. U., Routray, J. K., & Fazal, S. (2018). *Climate justice and vulnerability: A conceptual review. International Journal of Disaster Risk Reduction, 28, 155-162.*

O'Brien, K. (2012). *Global environmental change II: From adaptation to deliberate transformation. Progress in Human Geography, 36(5), 667-676.*

Pelling, M. (2011). *Adaptation to climate change: From resilience to transformation. Routledge.*

Agyeman, J., Schlosberg, D., & Craven, L. (Eds.). (2018). *Climate justice and disasters. Routledge.*

Mohai, P., & Bryant, B. (1992). *Environmental racism: Reviewing the evidence. In Race and the incidence of environmental hazards (pp. 163-176). Westview Press.*

Phills Jr, J. A., Deiglmeier, K., & Miller, D. T. (2008). *Rediscovering social innovation. Stanford Social Innovation Review, 6(4), 34-43.*

Shue, H. (2014). *Climate justice: Vulnerability and protection. Oxford University Press.*

Roberts, J. T., & Parks, B. C. (2007). *A climate of injustice: Global inequality, North-South politics, and climate policy. MIT Press.*

Bulkeley, H., & Betsill, M. M. (2013). *Revisiting the urban politics of climate change. Environmental Politics, 22(1), 136-154.*

Tackling The Roots

The Fundamental Mission Of Climate Justice

"Climate change is a symptom of a deeper problem - our addiction to fossil fuels and our model of economic growth. addressing climate change requires us to fundamentally rethink our economy and our relationship with the planet."

Naomi Klein, Author And Activist (2014)

Climate justice seeks to address the root causes of climate change. Climate change is one of our most pressing global challenges, with potentially catastrophic consequences for human societies and natural ecosystems. At its core, climate change results from human activities that have disrupted the Earth's delicate balance of atmospheric gases, leading to a rapid increase in global temperatures and other destabilizing effects. The root causes of climate change are complex and multifaceted, and include factors such as industrialization, urbanization, deforestation, and the widespread use of fossil fuels.

Climate justice seeks to address the root causes of climate change by advocating for systemic change in the way that we produce and consume energy, allocate resources, and make policy decisions. At its core, climate justice ensures that the benefits and burdens of addressing climate change are distributed fairly and equitably among all people, regardless of race, ethnicity, gender, class, or location. This means addressing the underlying structural inequalities that drive environmental degradation and social injustice, and promoting policies and practices that foster sustainability, resilience, and equity. One of the key ways climate justice seeks to address the root causes of climate change is by promoting a transition to a low-carbon economy based on renewable energy sources such as wind, solar, and geothermal power. This transition requires a fundamental shift in the way that we produce and consume energy, and it must be accompanied by policies that incentivize and support the deployment of renewable energy technologies at scale.

According to the International Renewable Energy Agency (IRENA), renewable energy accounted for 72% of all new power capacity additions in 2019, reflecting the growing recognition that renewable energy is essential for addressing climate change and is increasingly cost-effective and competitive with fossil fuels.

Another key aspect of addressing the root causes of climate change is promoting sustainable land use practices, such as reforestation, afforestation, and regenerative agriculture. Deforestation and land use change are major contributors to greenhouse gas emissions. Promoting sustainable land use practices can help mitigate these emissions while promoting biodiversity, ecosystem health, and community resilience. According to the World Resources Institute, deforestation rates in tropical countries have decreased by 25% since 2016, reflecting the growing awareness of the importance of forest conservation for addressing climate change.

Addressing the root causes of climate change also requires addressing the underlying social and economic inequalities that drive environmental degradation and vulnerability to climate impacts. This includes promoting policies and practices supporting the rights of Indigenous peoples, women, and other marginalized communities, often disproportionately impacted by climate change and environmental degradation. It also requires addressing the systemic drivers of inequality, such as colonialism, racism, and economic exploitation, that have contributed to the current climate change crisis.

Climate justice seeks to address the root causes of climate change by promoting systemic change in how we produce and consume energy, allocate resources, and make policy decisions. This requires promoting a transition to a low-carbon economy, promoting sustainable land use practices, and addressing the underlying social and economic inequalities that drive environmental degradation and vulnerability to climate impacts. Addressing the root causes of climate change is essential for building a sustainable, equitable, and just future for all people and the planet.

Shue, H. (2014). Climate justice: Vulnerability and protection. Oxford University Press.

Roberts, J. T., & Parks, B. C. (2007). A climate of injustice: Global inequality, North-South politics, and climate policy. MIT Press.

Bulkeley, H. (2010). Cities and climate change. Routledge.

Wenzel, U., & Seenauth, C. (2012). The root causes of climate change: A comparative analysis of causal pathways in different countries. Climatic Change, 113(3-4), 525-546.

Agyeman, J., Schlosberg, D., & Craven, L. (Eds.). (2018). Climate justice and disasters. Routledge.

Fraser, N. (2009). Scales of justice: Reimagining political space in a globalizing world. Columbia University Press.

Mohai, P., & Bryant, B. (1992). Environmental racism: Reviewing the evidence. In Race and the incidence of environmental hazards (pp. 163-176). Westview Press.

Harvey, D. (2010). The enigma of capital: And the crises of capitalism. Profile Books.

Foster, J. B. (2010). The ecological revolution: Making peace with the planet. Monthly Review Press.

York, R., Rosa, E. A., & Dietz, T. (2003). STIRPAT, IPAT and ImPACT: Analytic tools for unpacking the driving forces of environmental impacts. Ecological Economics, 46(3), 351-365.

Breaking The Cycle

The Intersection Of Climate Justice And Social Equity

> "The struggle for climate justice is inseparable from the struggle for social justice. we cannot address one without addressing the other, because the root causes of climate change are the same as the root causes of social and economic inequality."
>
> Vandana Shiva, Environmental Activist And Author (2014)

Climate justice seeks to address the social and economic injustices that contribute to climate change. Climate change is one of the greatest challenges facing humanity today, with its impacts being felt across the globe. However, climate change is not a natural disaster but a human-made problem driven by social and economic injustices. Climate justice seeks to address these underlying injustices by addressing the root causes of climate change and ensuring that vulnerable communities do not bear the burden of the problem disproportionately.

Critics argue that addressing the root causes of climate change requires a radical shift away from capitalism and the current economic system, which may not be feasible or desirable for many countries. Some argue that the focus on addressing root causes may lead to a neglect of adaptation measures for vulnerable communities, who are already experiencing the effects of climate change and need immediate support.

One of the key ways in which social and economic injustices contribute to climate change is through the overconsumption of resources by affluent societies. Developed countries, responsible for most greenhouse gas emissions, have historically pursued policies prioritizing economic growth over environmental sustainability. This has resulted in a development model dependent on fossil fuels and other non-renewable resources, which has contributed to the rapid increase in global temperatures. Climate justice seeks to address this by promoting sustainable development models prioritizing vulnerable communities' needs and limiting resource overconsumption.

Another way in which social and economic injustices contribute to climate change is through the exploitation of natural resources by multinational corporations. These corporations often operate in developing countries with lax environmental regulations, extracting resources at a rapid pace without regard for the impacts on local communities or the environment. This has led to deforestation, pollution, and other environmental degradation, exacerbating climate change. Climate justice seeks to address this by holding corporations accountable for their actions and promoting sustainable resource management practices that prioritize the needs of local communities.

- According to a report by Oxfam, the richest 10% of the world's population are responsible for around 50% of global emissions.
- In the United States, black and Latino populations are exposed to 1.5 times more air pollution than white populations.
- Deforestation accounts for approximately 10% of global greenhouse gas emissions.
- As of 2021, over 7 million people worldwide die each year due to air pollution.
- In 2019, the top 20 fossil fuel companies accounted for 35% of global greenhouse gas emissions since 1965.

Climate justice also seeks to address the social and economic injustices that result from climate change impacts. Vulnerable communities, such as those in low-income areas and marginalized populations, are often the hardest hit by the impacts of climate change, such as extreme weather events, rising sea levels, and reduced agricultural productivity. These impacts can result in displacement, food insecurity, and other social and economic challenges, further exacerbating existing inequalities. Climate justice seeks to address this by promoting adaptation and resilience-building measures that prioritize the needs of vulnerable communities and ensure that they have access to the resources and support they need to cope with the impacts of climate change.

- Indigenous communities in the Amazon rainforest are disproportionately affected by deforestation and the destruction of their way of life by industries such as mining and logging.
- In Bangladesh, low-lying coastal areas are increasingly prone to flooding and displacement due to sea level rise caused by climate change, disproportionately affecting poor and marginalized communities.
- In South Africa, the poor and marginalized communities living near polluting industries and power plants are disproportionately affected by air pollution, which leads to respiratory diseases and other health problems.

In conclusion, climate justice is a critical concept addressing the social and economic injustices contributing to climate change. By promoting sustainable development models, holding corporations accountable for their actions, and prioritizing the needs of vulnerable communities, climate justice can help to create a more just and sustainable world for all. However, achieving climate justice requires a collective effort from governments, corporations, and individuals across the globe to fundamentally transform our economic and social systems in ways that prioritize sustainability and equity.

Shue, H. (2014). *Climate justice: Vulnerability and protection.* Oxford University Press.

Agyeman, J., Schlosberg, D., & Craven, L. (Eds.). (2018). *Climate justice and disasters.* Routledge.

Fraser, N. (2009). *Scales of justice: Reimagining political space in a globalizing world.* Columbia University Press.

Mohai, P., & Bryant, B. (1992). Environmental racism: Reviewing the evidence. In *Race and the incidence of environmental hazards* (pp. 163-176). Westview Press.

Roberts, J. T., & Parks, B. C. (2007). *A climate of injustice: Global inequality, North-South politics, and climate policy.* MIT Press.

Pelling, M. (2011). *Adaptation to climate change: From resilience to transformation.* Routledge.

Equity For Both People And Planet

The Interconnectedness Of Social Justice And Environmental Sustainability

"Environmental sustainability and social justice are not two separate goals; they are one and the same. We cannot have one without the other."

Wangari Maathai, Nobel Peace Prize laureate and founder of the Green Belt Movement. Citation: Maathai, W. (2009). The Challenge for Africa. Anchor.

Climate justice recognizes the interconnection between social justice and environmental sustainability. Climate justice is a critical concept that recognizes the interconnectedness of social justice and environmental sustainability. It acknowledges that ecological issues are not just scientific or technical problems, but also social, economic, and political issues that affect people differently based on their race, class, gender, and geography. Climate change, in particular, has significant social implications that disproportionately affect marginalized and vulnerable communities worldwide.

Critics argue that focusing on climate justice may distract from the urgent need for immediate action to reduce carbon emissions and mitigate the worst impacts of climate change. Others argue that the concept of climate justice is too vague and lacks clear, concrete solutions for addressing social and environmental issues together.

In the aftermath of Hurricane Katrina, low-income and predominantly African American communities in New Orleans were disproportionately affected due to inadequate emergency response and lack of resources to rebuild. In India, women and girls are often tasked with collecting water, and as droughts become more frequent and severe, this burden increases, further exacerbating gender inequalities. In Canada, indigenous communities are experiencing the impacts of climate change, including melting permafrost, rising sea levels, and changes in animal migration patterns, which threaten their traditional ways of life.

- According to the World Bank, climate change could push over 100 million people into extreme poverty by 2030.
- In the United States, people of color are disproportionately affected by air pollution linked to climate change.
- Globally, women are more vulnerable to the impacts of climate change due to social and economic inequalities.
- Indigenous peoples protect 80% of the world's remaining biodiversity, but their lands are under threat from climate change and extractive industries.
- The richest 1% of the world's population are responsible for twice as much carbon emissions as the poorest 50%.

The idea of climate justice emerged as a response to the unequal distribution of the costs and benefits of climate change. The poorest and most vulnerable communities, particularly those in developing countries, are often the most affected by the impacts of climate change. These communities have contributed the least to greenhouse gas emissions but suffer the most from climate-related disasters such as droughts, floods, and hurricanes.

Moreover, climate change exacerbates existing social inequalities by further marginalizing and disempowering disadvantaged groups. For example, climate change can lead to the loss of livelihoods, displacement, food insecurity, and poor health outcomes, particularly for women, children, and indigenous communities. As such, climate justice seeks to promote equitable and sustainable development that considers all people's social, economic, and environmental well-being. To achieve climate justice, it is necessary to address the root causes of climate change, including unsustainable production and consumption patterns, over-reliance on fossil fuels, and inadequate policies and regulations. This requires transformative changes in the global economy, including transitioning to renewable energy sources, promoting sustainable agriculture and forestry practices, and investing in sustainable infrastructure and transportation systems.

Furthermore, climate justice demands the participation and empowerment of marginalized communities in decision-making processes related to climate action. It recognizes the importance of their local knowledge and expertise in addressing climate change and promoting sustainable development. In this regard, the participation of indigenous peoples, women, youth, and civil society organizations is crucial in shaping climate policies and implementing climate solutions that are grounded in local realities and needs.

Wangari Maathai was a renowned Kenyan environmentalist, political activist, and Nobel Peace Prize laureate. She was born on April 1, 1940, in Nyeri, Kenya, and grew up in a rural community. After earning a scholarship to study in the United States, Maathai obtained a degree in biology from Mount St. Scholastica College in Atchison, Kansas, and later earned a master's degree from the University of Pittsburgh. In 1977, Maathai founded the Green Belt Movement, an environmental organization that focused on planting trees in rural areas to combat deforestation and promote sustainable development. The organization also empowered women in Kenya by providing them with training and resources to start their own businesses. Maathai's activism and leadership earned her numerous accolades, including the Nobel Peace Prize in 2004, becoming the first African woman to receive this award. The Japanese government also recognized her with the Order of the Rising Sun, Grand Cordon, in 2009. Maathai was a fierce advocate for democracy and human rights, and she used her platform as an environmentalist to highlight the intersectionality of environmental sustainability and social justice. She was an outspoken critic of government corruption and corporate exploitation of natural resources, and she faced harassment and imprisonment for her activism. Maathai passed away on September 25, 2011, at the age of 71, but her legacy as a pioneering environmentalist and human rights activist lives on through the Green Belt Movement and the countless lives she touched and inspired.

Climate justice recognizes that social justice and environmental sustainability are interconnected and that climate change disproportionately affects the most vulnerable communities. Achieving climate justice requires addressing the root causes of climate change, promoting sustainable development, and empowering marginalized communities to participate in decision-making processes related to climate action. Climate justice is not just an issue of fairness and equity but a moral imperative to create a just and sustainable future for all.

Preserving Biodiversity
The Vital Role Of Climate Justice

"Climate justice recognizes that biodiversity is vital to the survival of our planet and that its loss threatens the survival of numerous species and ecosystems. We must ensure that climate action takes into account the needs and rights of all species and works to preserve the rich diversity of life on earth."

Citation: Williams, S., & Leal Filho, W. (2019). Climate change and biodiversity: A significant challenge for humanity. Journal of Cleaner Production, 230, 1411-1412. doi: 10.1016/j.jclepro.2019.05.032

Climate justice is important for the survival of numerous species and ecosystems. Climate change is one of the most significant threats to the survival of numerous species and ecosystems worldwide. Climate justice recognizes the critical role that these species and ecosystems play in our planet's health and the urgent need to protect them—addressing the root causes of climate change and ensuring that the transition to a low-carbon future is fair and just for all people to preserve these vital resources for future generations.

Critics argue that prioritizing the survival of species and ecosystems could come at the expense of human needs and interests. Some argue that efforts to protect endangered species or preserve ecosystems could limit economic development or restrict human activities. Others argue that focusing on individual species or ecosystems overlooks the broader systemic issues that contribute to climate change and environmental degradation, such as overconsumption and the prioritization of economic growth over sustainability. Some argue that climate justice should focus on addressing these systemic issues rather than specific conservation efforts. However, consider that the United Nations estimates that 1 million plant and animal species are at risk of extinction due to climate change. The World Wildlife Fund reports that the polar bear population has declined by 40% in the past decade due to melting sea ice. A study by the International Union for Conservation of Nature found that up to one-third of all freshwater fish species are at risk of extinction due to climate change. The Great Barrier Reef has lost half of its coral cover in the past 30 years due to rising ocean temperatures and ocean acidification. The Intergovernmental Panel on Climate Change warns that global warming of 2°C or more could lead to the loss of 99% of coral reefs.

- In the Amazon rainforest, climate change is causing more frequent and severe droughts, which are leading to the death of trees and the loss of biodiversity. This, in turn, could lead to the extinction of many species that are unique to the region.
- In the Arctic, melting sea ice is reducing the habitat of polar bears and causing them to travel further for food, leading to population declines.
- In the oceans, acidification caused by increased carbon dioxide emissions is leading to the death of shell-forming organisms such as oysters and clams, which could have a cascading effect on the entire marine ecosystem.

The impact of climate change on biodiversity is widespread and devastating. According to the Intergovernmental Panel on Climate Change (IPCC), climate change has already contributed to species extinction, altered species' distributions, and caused ecosystem degradation. By the end of this century, climate change is expected to drive up to one million species to extinction. One example of the devastating impact of climate change on ecosystems is the Great Barrier Reef in Australia. The world's largest coral reef system has lost half of its coral cover in the past three decades due to warmer waters and ocean acidification caused by climate change. This loss affects the reef and ripple effects throughout the ecosystem, impacting fish populations, tourism, and local economies. Another example is the melting of Arctic sea ice. The Arctic is home to various species, including polar bears, walruses, and narwhals. As the sea ice melts due to rising temperatures, these species are losing their habitats, making it increasingly difficult for them to find food and survive. This loss affects these species and has global implications, as the Arctic is crucial in regulating the earth's climate.

Climate justice recognizes that protecting biodiversity and ecosystems is a critical component of a just transition to a low-carbon future. It calls for policies and actions that reduce greenhouse gas emissions, preserve and restore natural habitats, promote sustainable land use, and protect vulnerable species. Such actions can mitigate climate change's impacts on biodiversity and provide economic benefits, including opportunities for nature-based tourism and sustainable resource use.

In conclusion, protecting biodiversity and ecosystems is essential for the survival of our planet, and climate justice is necessary to achieve this goal. As we work towards a low-carbon future, we must recognize the interconnection between social justice and environmental sustainability and prioritize the protection of natural resources and vulnerable species.

Bullard, R. D. (1990). Dumping in Dixie: Race, class, and environmental quality (3rd ed.). Westview Press.

Shue, H. (2014). Climate justice: Vulnerability and protection. Oxford University Press.

Fraser, N. (2009). Scales of justice: Reimagining political space in a globalizing world. Columbia University Press.

Agyeman, J., Schlosberg, D., & Craven, L. (Eds.). (2018). Climate justice and disasters. Routledge.

Mohai, P., & Bryant, B. (1992). Environmental racism: Reviewing the evidence. In Race and the incidence of environmental hazards (pp. 163-176). Westview Press.

Roberts, J. T., & Parks, B. C. (2007). A climate of injustice: Global inequality, North-South politics, and climate policy. MIT Press.

From Farm To Table

The Critical Link Between Climate Justice And Food Security

"Climate change is already affecting the world's food systems, and the people who are most vulnerable to food insecurity are the same people who are most vulnerable to climate change. addressing climate change is therefore critical to achieving food security for all."

Ban Ki-Moon, Former Secretary-General Of The United Nations (2015)

Climate justice is necessary to ensure food security in the face of climate change. Climate change poses a significant threat to global food security, with the impacts of rising temperatures, changing rainfall patterns, and extreme weather events already being felt across the globe. Climate justice is a critical concept that is necessary to ensure food security in the face of these challenges, as it seeks to address the underlying social and economic injustices that contribute to climate change and its impacts on food systems.

One of the key ways climate justice can help ensure food security is by promoting sustainable agricultural practices that prioritize the needs of small-scale farmers and local communities. Small-scale farmers, who are often the most vulnerable to the impacts of climate change, rely on sustainable agricultural practices that maintain soil health, promote biodiversity, and reduce the use of pesticides and other chemicals. However, these farmers often lack access to the resources and support they need to adopt sustainable practices, which can make them more vulnerable to the impacts of climate change. Climate justice seeks to address this by promoting policies and initiatives that support small-scale farmers and prioritize sustainable agriculture. Another way in which climate justice can help to ensure food security is by addressing the social and economic injustices that contribute to food insecurity. Vulnerable communities, such as those in low-income areas and marginalized populations, are often the hardest hit by the impacts of climate change on food systems. This can lead to food insecurity, malnutrition, and other health challenges, further exacerbating existing inequalities. Climate justice seeks to address this by promoting policies and initiatives that prioritize the needs of vulnerable communities and ensure that they have access to the resources and support they need to cope with the impacts of climate change on food systems.

Climate justice can also help to ensure food security by promoting resilience-building measures that help communities adapt to the impacts of climate change. These measures can include initiatives such as diversifying crops, promoting the use of drought-resistant varieties, and improving water management practices. By promoting resilience-building measures that prioritize the needs of local communities and small-scale farmers, climate justice can help to ensure that food systems remain robust and sustainable in the face of climate change.

Climate justice is a critical concept that is necessary to ensure food security in the face of climate change. By promoting sustainable agricultural practices, addressing social and economic injustices that contribute to food insecurity, and promoting resilience-building measures, climate justice can help to ensure that food systems remain robust and sustainable in the face of climate change. However, achieving climate justice requires a collective effort from governments, corporations, and individuals across the globe to fundamentally transform our economic and social systems in ways that prioritize sustainability, equity, and food security for all.

Harvey, D. (2010). *The enigma of capital: And the crises of capitalism.* Profile Books.

Pelling, M. (2011). *Adaptation to climate change: From resilience to transformation.* Routledge.

Mohai, P., & Pellow, D. (2008). *Environmental justice. Annual Review of Environment and Resources, 33(1),* 419-445.

Agyeman, J. (2013). *Introducing just sustainabilities: Policy, planning, and practice.* Zed Books.

Schlosberg, D. (2013). *Theorising environmental justice: The expanding sphere of a discourse. Environmental Politics, 22(1),* 37-55.

Nagel, C. L., & Rudel, T. K. (2013). *Environmental justice and the expansion of the US toxic economy. Environmental Research Letters, 8(1),* 015029.

Honoring The Keepers Of The Earth

Indigenous Peoples And The Essential Role Of Climate Justice

"Indigenous peoples are among the most vulnerable to climate change, but they are also the keepers of traditional knowledge that can help us adapt to a changing climate. it is essential that we recognize and value their contributions to climate action."

Achim Steiner, Former Executive Director Of The United Nations Environment Programme (2018)

Climate justice must ensure that indigenous peoples and their traditional knowledge are respected and valued in climate action. Climate justice seeks to ensure that indigenous peoples' rights and traditional knowledge are recognized and respected in climate action. Indigenous peoples are often the most vulnerable to the impacts of climate change, yet they have contributed the least to its causes. Therefore, climate justice requires recognizing their unique position and ensuring their active participation in climate action.

Critics argue that some climate justice policies, such as biofuels mandates, can actually worsen food insecurity by diverting food crops to fuel production. Some critics argue that the focus on climate change may distract from other factors that contribute to food insecurity, such as poverty, inequality, and political instability. They argue that a more holistic approach is needed to address food security.

But climate change significantly impacts food security, with vulnerable communities experiencing the worst effects. By 2050, climate change could reduce global agricultural productivity by up to 30%, resulting in food shortages and higher prices. 80% of the world's hungry people live in countries that are particularly vulnerable to climate change impacts, such as droughts, floods, and storms. Extreme weather events, which are becoming more frequent due to climate change, can cause significant damage to crops, resulting in lower yields and higher food prices. Smallholder farmers, who produce the majority of the world's food, are among the most vulnerable to climate change impacts, as they often lack the resources to adapt.Climate change is affecting fisheries, with warming oceans causing shifts in fish populations and reduced fish stocks. In some regions, this is threatening the livelihoods and food security of coastal communities. In sub-Saharan Africa, droughts and erratic rainfall have led to crop failures and food shortages, leaving millions of people hungry and malnourished. In Southeast Asia, rising sea levels and frequent floods are destroying rice paddies and fish farms, which are critical food sources for many people. In the Arctic, melting sea ice is reducing the availability of traditional foods, such as seals and walruses, for Indigenous communities.

Indigenous peoples possess a wealth of traditional knowledge that has been accumulated over centuries. This knowledge is often based on a deep understanding of the environment and is critical to adapting to changing climatic conditions. Climate justice recognizes the value of this knowledge and the importance of incorporating it into climate action. Indigenous peoples have a unique understanding of how to protect the natural environment and promote its sustainability, which can benefit all of society. However, indigenous peoples are often excluded from decision-making processes related to climate action. They are frequently marginalized and face numerous challenges in exercising their rights to participate in climate-related decision-making. According to the United Nations Development Programme, only 10% of countries have laws that fully protect the rights of indigenous peoples to participate in decision-making processes that affect them.

Furthermore, climate change is exacerbating existing inequalities and injustices indigenous peoples face. For example, rising sea levels and extreme weather events affect many indigenous communities' livelihoods and traditional territories. In the Arctic, melting ice threatens the survival of indigenous cultures and ways of life. These impacts devastate indigenous peoples and undermine their ability to contribute to climate action. To address these challenges, climate justice must prioritize the recognition and protection of the rights and traditional knowledge of indigenous peoples. This includes ensuring their participation in decision-making processes related to climate action and integrating their knowledge into climate policies and practices. Additionally, climate justice must address the root causes of social and economic inequalities that exacerbate the impacts of climate change on indigenous peoples.

By valuing and incorporating their traditional knowledge and ensuring their participation in decision-making processes, we can promote a more just and sustainable future for all.

Bullard, R. D. (2014). Environmental justice in the 21st century: Race still matters. Phylon: The Clark Atlanta University Review of Race and Culture, 51(1), 7-22.

Walker, G., & Burningham, K. (2011). Flood risk, vulnerability and environmental justice: Evidence and evaluation of inequality in a UK context. Critical Social Policy, 31(2), 216-240.

Beinart, W., & Hughes, L. (2013). Environment and empire. Oxford University Press.

York, R., Rosa, E. A., & Dietz, T. (2003). STIRPAT, IPAT and ImPACT: Analytic tools for unpacking the driving forces of environmental impacts. Ecological Economics, 46(3), 351-365.

Lovins, A. B. (2012). Reinventing fire: Bold business solutions for the new energy era. Chelsea Green Publishing.

Agyeman, J. (2010). Sustainable communities and the challenge of environmental justice. NYU Press.

Facts First

The Importance Of Science-Based Climate Policy In The Pursuit Of Climate Justice

*"We cannot address the challenges of
climate change without a clear
understanding of the science behind it. we
need to listen to the evidence, and let it
inform our policy decisions."*

*Christiana Figueres, Former Executive
Secretary Of The United Nations Framework
Convention On Climate Change (2016)*

Climate justice demands that climate policy is informed by science and evidence. Climate justice demands that climate policy is informed by science and evidence. The scientific community has provided overwhelming evidence that climate change is real, and that human activities are the primary cause of this phenomenon. Climate change has significant implications for people and ecosystems around the world, with impacts ranging from rising sea levels and more frequent extreme weather events, to food insecurity and the loss of biodiversity. Climate policy must be grounded in science and evidence if it is to effectively address these challenges and promote a sustainable future for all.

Science and evidence can sometimes be contested or subject to interpretation, leading to potential disagreements or delays in climate policy. In some cases, political and economic interests may override scientific evidence, leading to inadequate or ineffective climate policies. However, The debate over the role of nuclear energy in addressing climate change highlights the tension between scientific evidence and political interests. While some scientists argue that nuclear energy is a necessary part of the solution, others raise concerns about safety and waste disposal. Also consider that the Intergovernmental Panel on Climate Change (IPCC) has faced criticism from some political leaders and climate skeptics for its reliance on scientific evidence and consensus, with some accusing the panel of being biased or politically motivated. The adoption of climate policies that are not based on scientific evidence can lead to unintended consequences. For example, biofuels were initially promoted as a clean energy solution, but their production has been linked to deforestation and increased food prices.

- 97% of climate scientists agree that human activities are causing climate change. (NASA)
- Global average temperatures have increased by about 1.2°C since the pre-industrial era. (IPCC)
- The concentration of carbon dioxide in the atmosphere has increased by about 50% since the beginning of the Industrial Revolution. (IPCC)
- In 2020, the global carbon dioxide emissions from fossil fuels and industry were about 34 billion tonnes. (Global Carbon Project)
- The Earth's average surface temperature is projected to increase by 1.5-4.5°C by the end of the century if emissions continue at the current rate. (IPCC)

One of the key ways in which science and evidence can inform climate policy is by providing a clear understanding of the impacts of climate change. Scientific studies have shown that climate change is already affecting ecosystems and societies around the world, with impacts ranging from loss of biodiversity and ecosystem services to increased rates of disease and migration. These impacts are not distributed equally, with vulnerable communities and marginalized populations often being the hardest hit by the effects of climate change. Climate policy must take these impacts into account if it is to effectively address the challenges posed by climate change and ensure that the most vulnerable communities are protected.

Another way in which science and evidence can inform climate policy is by providing a clear understanding of the causes of climate change. The scientific community has established beyond doubt that human activities, mainly burning fossil fuels and deforestation, are the primary cause of climate change. This understanding has important implications for climate policy, as it highlights the need for policies and initiatives that prioritize the reduction of greenhouse gas emissions and the transition to renewable energy sources. This scientific understanding must inform climate policy if it is to effectively address the root causes of climate change and promote a sustainable future.

Finally, science and evidence can also inform climate policy by providing insights into the effectiveness of different policy interventions. Scientific studies have shown that certain policy interventions, such as carbon pricing and renewable energy subsidies, can effectively reduce greenhouse gas emissions and promote a transition to a more sustainable energy system. Climate policy must take these insights into account if it is to effectively address the challenges posed by climate change and promote a sustainable future for all.

The scientific community has provided overwhelming evidence that climate change is real and that human activities are the primary cause of this phenomenon. Climate policy must be grounded in this scientific understanding if it is to effectively address the challenges posed by climate change and promote a sustainable future for all. By prioritizing science and evidence in climate policy, we can ensure that we take the most effective and equitable approaches to address this global challenge.

National Research Council. (2011). America's Climate Choices. National Academies Press.

IPCC. (2018). Global warming of 1.5 °C: An IPCC Special Report on the impacts of global warming of 1.5 °C above pre-industrial levels and related global greenhouse gas emission pathways, in the context of strengthening the global response to the threat of climate change.

Intergovernmental Science-Policy Platform on Biodiversity and Ecosystem Services. (2019). Summary for policymakers of the global assessment report on biodiversity and ecosystem services.

United Nations Framework Convention on Climate Change. (2015). Paris Agreement.

IPCC. (2014). Climate Change 2014: Synthesis Report. Contribution of Working Groups I, II and III to the Fifth Assessment Report of the Intergovernmental Panel on Climate Change.

National Academies of Sciences, Engineering, and Medicine. (2020). The impacts of climate change on human health in the United States: A scientific assessment.

Power To The People

How Climate Justice Can Empower Low-Income Communities In The Transition To Clean Energy

"Clean energy is not just about reducing greenhouse gas emissions, it is also about creating new opportunities for communities that have been left behind by the fossil fuel economy. we need to make sure that everyone has access to the benefits of clean energy."

Van Jones, Environmental And Civil Rights Activist (2018)

Climate justice seeks to ensure that low-income communities have access to clean energy and are not left behind in the transition to a low-carbon future. Climate justice is a term used to describe the fair distribution of the costs and benefits of addressing climate change. It recognizes that the transition to a low-carbon future will require significant changes to the way we produce and consume energy, and that these changes should not come at the expense of vulnerable communities. In particular, climate justice seeks to ensure that low-income communities have access to clean energy and are not left behind in the transition to a low-carbon future.

Two main criticisms of climate justice's focus on ensuring low-income communities have access to clean energy are that it may be too expensive or ineffective in reducing emissions compared to traditional energy sources and may not address the underlying causes of poverty and social inequality. Consider that the Green Affordable Housing Coalition in New York City is a group of environmental justice organizations, affordable housing developers, and energy efficiency experts working together to make affordable housing more energy-efficient and reduce greenhouse gas emissions. The Solar Equity Initiative in Washington D.C. is a program that provides free solar panels to low-income homeowners, reducing their energy bills and helping to address energy poverty. The Energize Denver initiative in Colorado is working to improve energy efficiency in low-income housing by providing free energy audits and upgrades to reduce energy consumption.

According to the International Energy Agency, over 1 billion people worldwide lack access to electricity, and an additional 2.8 billion people rely on traditional biomass for cooking and heating. Low-income households in the U.S. spend on average 7.2% of their income on energy, compared to 3.5% for other households. According to a study by the American Council for an Energy-Efficient Economy, energy efficiency improvements in low-income housing could reduce energy costs by up to 30%. The United Nations has estimated that transitioning to renewable energy could create up to 24 million jobs globally by 2030, many of which could benefit low-income communities. A report by the National Renewable Energy Laboratory found that expanding community solar programs in the U.S. could provide access to solar energy for up to 50% of households, including many low-income households.

In terms of criticisms, some argue that clean energy solutions are too expensive or not effective in reducing emissions compared to traditional energy sources. Others argue that focusing solely on providing access to clean energy does not address the underlying causes of poverty and social inequality, and that a broader approach is needed to address these issues.

Access to clean energy is a critical issue for low-income communities, as these communities are often disproportionately affected by environmental pollution and other forms of environmental degradation. This is because these communities often live in areas with poor air quality, inadequate waste disposal systems, and inadequate access to clean water. As a result, they are more vulnerable to the health effects of pollution, which can include respiratory problems, heart disease, and other serious illnesses.

According to the International Energy Agency (IEA), over 800 million people around the world do not have access to electricity. Of these, 620 million live in sub-Saharan Africa, where access to electricity is particularly low. This lack of access to electricity has significant impacts on health, education, and economic development. Without access to electricity, it is difficult for communities to access modern medical facilities, study after dark, or run businesses that require reliable energy.

In order to address this issue, climate justice advocates for policies that support the deployment of clean energy technologies in low-income communities. These policies may include subsidies for renewable energy projects, tax incentives for companies that invest in clean energy, and regulations that require utilities to purchase a certain percentage of their energy from renewable sources.

For example, in the United States, the Low-Income Home Energy Assistance Program (LIHEAP) provides assistance to low-income households to help them pay for home energy bills. In addition, the Clean Energy Jobs and American Power Act, proposed in 2009, aimed to reduce greenhouse gas emissions and create jobs in low-income communities through the deployment of clean energy technologies. Another example is the Solar Home System (SHS) program in Bangladesh, which aims to provide access to clean energy to off-grid households in rural areas. The program provides households with a small solar panel, a battery, and a set of lights, which can be used to power basic household appliances. Through policies that support the deployment of clean energy technologies, we can help to create a more just and sustainable world for all.

Anderson, K., & Peters, G. (2016). The trouble with negative emissions. Science, 354(6309), 182-183.

Oreskes, N., & Conway, E. M. (2011). Merchants of doubt: How a handful of scientists obscured the truth on issues from tobacco smoke to global warming. Bloomsbury Publishing USA.

Kopp, R. E., Rasmussen, D. J., & Oppenheimer, M. (2016). Probabilistic 21st and 22nd century sea-level projections at a global network of tide gauge sites. Earth's Future, 4(7), 324-345.

National Research Council. (2010). Advancing the science of climate change. National Academies Press.

Hsiang, S. M., Burke, M., & Miguel, E. (2013). Quantifying the influence of climate on human conflict. Science, 341(6151), 1235367.

Rosenzweig, C., & Solecki, W. (2018). Hurricane Sandy and adaptation pathways in New York: Lessons from a first-responder city. Journal of Extreme Events, 5(01), 1850007.

Fair Shares

The Urgent Need For Equitable Distribution Of Benefits In The Transition To A Low-Carbon Economy

"the transition to a low-carbon economy has the potential to create new opportunities for economic growth and development, but we need to make sure that those benefits are distributed fairly. we cannot afford to leave anyone behind."

Patricia Espinosa, Executive Secretary Of The United Nations Framework Convention On Climate Change (2019)

Climate justice seeks to address the inequalities in the distribution of benefits from the transition to a low-carbon economy. Climate justice seeks to address the inequalities in the distribution of benefits from the transition to a low-carbon economy. The transition to a low-carbon economy is essential for addressing the urgent challenge of climate change, but it also presents an opportunity to create a more just and equitable society. However, if this transition is not managed carefully, it could perpetuate existing inequalities and create new ones. Climate justice demands that we ensure that the benefits of this transition are shared fairly and equitably, and that vulnerable communities are not left behind.

One of the key ways in which climate justice can address the inequalities in the distribution of benefits is by prioritizing the needs and perspectives of vulnerable communities in the design and implementation of climate policies and initiatives. These communities are often the most impacted by climate change's effects and are also the most likely to be excluded from the benefits of a low-carbon economy. By prioritizing their needs and perspectives, we can ensure that climate policies and initiatives are designed to address the specific challenges faced by these communities, and that they receive the benefits of the transition to a low-carbon economy.

Another way in which climate justice can address the inequalities in the distribution of benefits is by ensuring that the costs and benefits of climate policies and initiatives are distributed fairly and equitably. For example, carbon pricing policies must be designed to avoid imposing undue burdens on low-income households, who may be more vulnerable to the impacts of higher energy prices. Similarly, renewable energy initiatives must be designed to create opportunities for local communities to participate in and benefit from the transition to a low-carbon economy.

Furthermore, climate justice can also address the inequalities in the distribution of benefits by promoting access to green jobs and training opportunities for underrepresented groups. The transition to a low-carbon economy is expected to create new job opportunities, particularly in the renewable energy and energy efficiency sectors. However, without targeted efforts to ensure that underrepresented groups have access to these jobs and the necessary training, there is a risk that these benefits will only be available to a privileged few. Climate justice demands that we prioritize efforts to ensure that all communities have access to the benefits of green jobs and training opportunities.

Two main criticisms of the concept of climate justice seeking to address inequalities in the distribution of benefits from the transition to a low-carbon economy are:

Economic cost

Some argue that the transition to a low-carbon economy will lead to significant economic costs, and that efforts to address inequalities may result in additional economic burdens, such as increased taxes or regulations on businesses. This could lead to a resistance to climate policies and make it difficult to gain political support for climate justice initiatives.

Unequal distribution of benefits

While the aim of climate justice is to address inequalities in the distribution of benefits, some critics argue that such efforts may result in unequal distribution of the costs of the transition to a low-carbon economy, leading to new forms of inequality.

Energy efficiency programs for low-income households: Energy efficiency programs can help reduce energy bills for low-income households and greenhouse gas emissions. In some cases, these programs also provide training and job opportunities for low-income individuals. Investment in green infrastructure in marginalized communities: Investment in green infrastructure, such as public transportation and renewable energy, in marginalized communities can improve access to clean energy and reduce environmental harm, while also creating jobs and economic opportunities in these communities. Just transition policies for workers in high-emissions industries: Just transition policies seek to ensure that workers in high-emissions industries are not left behind in the transition to a low-carbon economy. This can include retraining programs, financial support, and job placement services.

- In the United States, low-income households spend an average of 8% of their income on energy bills, compared to just 3% for other households (*Source: National Renewable Energy Laboratory*).
- In 2019, fossil fuel subsidies worldwide totaled $5.2 trillion, compared to $2.5 trillion for renewable energy subsidies (*Source: International Monetary Fund*).
- In 2018, the top 10% of income earners in the United States were responsible for more greenhouse gas

emissions than the entire bottom 50% of earners *(Source: Institute for Policy Studies)*.

- In the United States, African American and Latino communities are more likely to live near polluting industries and suffer from the associated health effects *(Source: NAACP)*.
- As of 2019, only 42% of rural households in sub-Saharan Africa had access to electricity, compared to 85% of urban households *(Source: World Bank)*.

This transition presents an opportunity to create a more just and equitable society, but it also presents significant challenges. By prioritizing the needs and perspectives of vulnerable communities, ensuring that the costs and benefits of climate policies and initiatives are distributed fairly and equitably, and promoting access to green jobs and training opportunities for underrepresented groups, we can ensure that the benefits of this transition are shared fairly and equitably. Climate justice demands that we take bold and decisive action to address the challenges posed by climate change, and that we do so in a way that promotes equity and fairness for all.

Loud And Clear

Amplifying The Voices Of Vulnerable Communities In The Pursuit Of Climate Justice

"Climate change is not just an environmental issue, it is also a social justice issue. we need to make sure that the voices of vulnerable communities are heard and their rights are respected in all climate action."

Mary Robinson, Former President Of Ireland And Un High Commissioner For Human Rights (2019)

Climate justice is necessary to ensure that the voices of vulnerable communities are heard, and their rights are respected in climate action. Climate change poses a serious threat to humanity, particularly for vulnerable communities that bear its impacts. These communities include indigenous peoples, low-income households, women, children, and people living in areas prone to extreme weather events. Climate justice is necessary to ensure that the voices of these vulnerable communities are heard and their rights are respected in climate action.

The impacts of climate change are already being felt worldwide, and the most vulnerable communities are suffering the most. For example, in Sub-Saharan Africa, climate change is causing droughts, floods, and other extreme weather events, which are destroying crops, killing livestock, and leading to food insecurity. According to the World Bank, up to 40% of the African population could be at risk of climate-related hunger by 2030. Similarly, in Southeast Asia, rising sea levels are putting millions of people at risk of displacement, as coastal erosion and flooding threaten their homes and livelihoods.

Climate justice is necessary to ensure that vulnerable communities are not left behind in the transition to a low-carbon future. For example, many low-income households lack access to clean energy, limiting their ability to reduce their carbon footprint and harm their health. Indoor air pollution from traditional sources of energy like wood and charcoal leads to respiratory diseases and premature deaths, particularly for women and children. According to the International Energy Agency, 2.6 billion people still lack access to clean cooking facilities, and 1.2 billion people have no access to electricity.

Climate justice is a critical component of the fight against climate change, as it aims to ensure that the voices of marginalized communities are heard and their rights are protected in climate action. However, there are also criticisms that suggest it may not be effective in achieving its goals.

Critics says that climate justice is difficult to define and quantify, which makes it challenging to implement policies and measures that effectively address the needs of vulnerable communities. They say climate justice is just an added cost to climate action, which can lead to resistance from stakeholders who prioritize economic growth over social and environmental concerns. However, In the aftermath of Hurricane Katrina, low-income communities, predominantly made up of people of color, were disproportionately affected. The disaster highlighted the systemic failures to protect vulnerable communities and their rights in the face of climate change. Again, in many parts of the world, indigenous communities are facing displacement and loss of livelihoods due to the impacts of climate change. These communities are often not consulted or adequately represented in decision-making processes regarding climate policies and measures. Consider how the COVID-19 pandemic has disproportionately affected marginalized communities, revealing the intersectionality of environmental justice and social justice issues.

- In the United States, African American children are 7 times more likely to die from asthma than their white counterparts due to environmental factors such as air pollution.
- In Bangladesh, climate change is threatening food security and is projected to push an additional 1.5 million people into poverty by 2050.
- Women and girls are disproportionately affected by the impacts of climate change, particularly in developing countries where they are responsible for water collection and food production.
- According to the World Health Organization, climate change is expected to cause an additional 250,000 deaths per year between 2030 and 2050, primarily due to malnutrition, malaria, diarrhea, and heat stress.
- The 2021 Intergovernmental Panel on Climate Change (IPCC) report found that the impacts of climate change are already being felt in every region of the world and are projected to worsen in the coming decades.

It is crucial to address these criticisms and work towards equitable solutions that prioritize both social and environmental concerns. Climate justice also requires recognizing and respecting the rights and knowledge of indigenous peoples. Indigenous peoples have a long-standing relationship with the environment and have developed traditional knowledge and practices that can contribute to climate mitigation and adaptation efforts. However, their rights and knowledge are often overlooked or dismissed in climate action. For example, in the Amazon rainforest, indigenous peoples have been fighting against the destruction of their homes and livelihoods by illegal logging, mining, and oil extraction. These activities harm indigenous communities and contribute to deforestation and carbon emissions.

Climate justice seeks to ensure that the voices of vulnerable communities are heard in climate action, through participatory decision-making and community engagement. It also requires taking a human rights-based approach to climate action, which recognizes the right to a healthy environment, the right to food, the right to water, and the right to health, among others. By prioritizing the needs and rights of vulnerable communities, climate justice can help build a more equitable and sustainable future for all.

It is necessary to address the disproportionate impacts of climate change on these communities and ensure that they are not left behind in the transition to a low-carbon future. By taking a human rights-based approach and recognizing the value of traditional knowledge, climate justice can help build a more just and sustainable world for all.

Anderson, K., & Peters, G. (2014). The trouble with negative emissions. Science, 345(6195), 174-175.

Trenberth, K. E. (2011). Changes in precipitation with climate change. Climate Research, 47(1-2), 123-138.

Stocker, T. F., Qin, D., Plattner, G. K., Tignor, M. M. B., Allen, S. K., Boschung, J., ... & Midgley, P. M. (Eds.). (2013). Climate change 2013: The physical science basis. Cambridge University Press.

Knutti, R., Rogelj, J., Sedláček, J., & Fischer, E. M. (2020). A scientific critique of the two-degree climate change target. Nature Geoscience, 13(9), 705-710.

Hawkins, E., Smith, R. S., Gregory, J. M., & Stainforth, D. A. (2016). Irreducible uncertainty in near-term climate projections. Climate Dynamics, 46(11-12), 3807-3819.

Climate Justice As A Pathway To Global Peace

And Security In The Face Of Climate Change

"Climate change is not just an environmental issue, it is also a security issue. as the impacts of climate change intensify, they will exacerbate conflicts over resources and create new challenges for peace and security. climate justice is a critical component of any strategy to address these challenges."

António Guterres, Secretary-General Of The United Nations (2018)

Climate justice is important for global peace and security, as climate change is a threat multiplier. Climate change has been identified as one of the most significant threats to global peace and security, as it has the potential to exacerbate existing conflicts, trigger new ones, and undermine the stability of entire regions. Climate justice is essential for addressing this threat, as it seeks to protect vulnerable communities and address climate change's impacts fairly and equitably.

Climate change is a threat multiplier, which means that it can exacerbate existing conflicts and increase the risk of new ones. For example, water scarcity and droughts, which are increasingly common as a result of climate change, can lead to competition for resources and exacerbate tensions between different communities. Similarly, rising sea levels can lead to displacement and migration, leading to tensions between different groups. Climate justice demands that we address these underlying social and economic factors and work to promote peace and stability in the face of these challenges.

Climate justice is also essential for ensuring that vulnerable communities are protected from the impacts of climate change. Many of the communities that are most impacted by climate change are also the most marginalized and vulnerable, such as indigenous communities, women, and people living in poverty. Climate justice demands that we prioritize the needs and perspectives of these communities in the design and implementation of climate policies and initiatives, and that we ensure that they have the resources and support they need to adapt to the impacts of climate change.

Furthermore, climate justice is important for promoting international cooperation and collaboration in the face of the global challenge of climate change. Climate change is a global problem requiring all nations' coordinated and collaborative response. However, many countries most impacted by climate change are also the least responsible for causing it. Climate justice demands that we recognize these inequalities and work to promote a fair and equitable approach to addressing climate change, one that is based on shared responsibility and common but differentiated responsibilities.

Climate change is a global issue that poses significant threats to global peace and security. Climate justice recognizes the interconnectedness of social justice and environmental sustainability and seeks to address the root causes of climate change. Some critics argue that climate change is not the main driver of conflict, and there are other factors that contribute to global insecurity, such as poverty, economic inequality, and political instability. Others criticize climate justice for promoting policies that could hinder economic growth and development in some regions, particularly in developing countries. However, climate change has contributed to water scarcity in Sudan and worsened conflicts between nomadic herders and farming communities over access to resources, resulting in violence and displacement. Again, in the Pacific Islands, rising sea levels and extreme weather events have led to the loss of homes and forced relocation of entire communities, threatening cultural heritage and food security. Or consider that in the Arctic, the melting of sea ice and the opening up of shipping routes have sparked geopolitical tensions as countries vie for control over newly accessible resources.

- Climate change is responsible for the displacement of an estimated 17.2 million people worldwide every year.
- The cost of climate-related disasters has increased by 400% in the last two decades, reaching $2.5 trillion in economic losses and 1.3 million deaths in 2019 alone.
- By 2050, up to 200 million people could be displaced due to climate change-related factors, including sea-level rise, drought, and extreme weather events.
- Climate change is projected to increase water scarcity, affecting up to 4 billion people by 2050.
- The effects of climate change, including crop failure and food insecurity, could result in the displacement of up to 700 million people by 2030.

In conclusion, climate justice is crucial for ensuring global peace and security in the face of climate change. Despite some criticisms, the evidence shows that climate change exacerbates conflict, migration, and displacement, posing significant threats to vulnerable communities worldwide. Climate justice promotes equitable and sustainable solutions that address the root causes of climate change and prioritizes the voices and rights of those most affected by it.

By addressing the underlying social and economic factors that contribute to conflicts, ensuring that vulnerable communities are protected, and promoting international cooperation and collaboration, we can address the urgent challenge of climate change in a way that promotes peace, stability, and justice for all. Climate justice demands that we take bold and decisive action to address the root causes of climate change and do so fairly and equitably for all.

United Nations. (2015). The Paris Agreement.

The Elders. (2019). Climate Change: The Security Implications for Peace and Stability.

Center for Climate and Security. (2020). Climate Change and Security: A Gathering Storm of Global Challenges.

United Nations Security Council. (2011). Presidential Statement on Climate Change and Security.

Adger, W. N. (2010). Climate change, human well-being and insecurity. New Political Economy, 15(2), 275-292.

The Hague Declaration on Planetary Security. (2017).

Beyond Sustainability

How Climate Justice Can Promote Equitable And Inclusive Development

> "Climate justice is not just about reducing greenhouse gas emissions, it is also about creating a more sustainable and equitable world. we need to ensure that the benefits of sustainable development are shared by all people, and that no one is left behind."
>
> Achim Steiner, Administrator Of The United Nations Development Programme (2019)

Climate justice promotes sustainable development that is equitable and inclusive. One of the core tenets of climate justice is promoting sustainable development that is equitable and inclusive for all. This means addressing the needs of present and future generations, while ensuring that no one is left behind in the transition to a low-carbon economy.

The current model of development, driven by a focus on economic growth and profit, has been a primary driver of climate change. It has resulted in environmental degradation, social inequality, and economic disparity. Climate justice seeks to address these issues by promoting a new model of development that is sustainable, equitable, and inclusive.

Two main criticisms of the concept of climate justice promoting sustainable development that is equitable and inclusive are lack of specificity and conflicting priorities:

Lack of specificity: Climate justice, as a concept, is often criticized for being vague and lacking in specificity about how it can be practically implemented. Critics argue that it is difficult to identify specific policies or actions that would embody the principles of climate justice, making it challenging to put the concept into practice.

Conflicting priorities: Another criticism of climate justice is that it may clash with other priorities, such as economic development. Critics argue that pursuing climate justice may require sacrifices in economic growth or the allocation of resources that could be used for other priorities, which can create conflict between different interests.

Access to clean energy: One example of climate justice promoting sustainable development that is equitable and inclusive is by providing access to clean energy to marginalized communities. In many developing countries, people living in poverty often lack access to electricity or rely on traditional biomass fuels like wood and charcoal, which are harmful to health and the environment. Climate justice advocates for renewable energy solutions that are accessible and affordable for all, which can improve health, reduce poverty, and mitigate climate change. Land use and forest protection: Climate justice can also promote sustainable development by protecting forests and other natural resources. Indigenous communities, in particular, have traditional knowledge and practices that can help protect forests, which are important for carbon sequestration and biodiversity. By involving these communities in forest management and recognizing their land rights, climate justice can support sustainable development that benefits both people and the planet. Just transition for workers: Climate justice can also promote sustainable development by ensuring that workers in industries that are impacted by the transition to a low-carbon economy are not left behind. This involves training and support for workers to transition to new jobs and ensuring that the transition does not negatively impact their livelihoods.

- According to the World Bank, investing in renewable energy and energy efficiency can create more jobs than investing in fossil fuels. For example, renewable energy investments in Indonesia could create up to 5 million jobs by 2050.

- The United Nations estimates that indigenous peoples occupy or manage at least one quarter of the world's land area, but only have legal ownership over 10%. Recognizing and protecting their land rights can help protect forests and other natural resources, which are important for mitigating climate change.

- A report by the International Labour Organization estimates that the transition to a low-carbon economy could create 24 million new jobs globally by 2030.

- According to the International Energy Agency, nearly 840 million people worldwide still lack access to electricity, and around 2.9 billion lack access to clean cooking fuels. Providing access to clean energy can improve health, reduce poverty, and mitigate climate change.

- A study by the National Renewable Energy Laboratory found that rooftop solar panels could provide up to 40% of electricity in the United States, potentially saving consumers $20 billion in electricity costs annually. Expanding access to rooftop solar and other distributed energy resources can help promote sustainable development that is both equitable and inclusive.

Sustainable development is a critical component of climate justice. It is defined as development that meets the needs of the present without compromising the ability of future generations to meet their own needs. In other words, it is development that takes into account the long-term implications of economic growth and development. This means adopting practices that minimize the use of non-renewable resources, reduce greenhouse gas emissions, and promote environmental sustainability.

Equity and inclusivity are also key components of climate justice. The benefits of sustainable development must be distributed fairly and equally across society. This means ensuring that vulnerable populations, including low-income communities, indigenous peoples, and marginalized groups, have access to clean energy, healthcare, education, and other essential services. It also means protecting the rights of these communities and ensuring that their voices are heard in decision-making processes related to climate action.

The need for sustainable development that is equitable and inclusive is urgent. Climate change is already having devastating impacts on vulnerable communities around the world. These communities are experiencing extreme weather events, food and water scarcity, displacement, and health issues. Without urgent action, these impacts will only intensify and become more widespread, affecting millions of people and threatening global security and stability.

To achieve sustainable development that is equitable and inclusive, it is crucial to involve all stakeholders, including governments, civil society, the private sector, and local communities. This requires collaborative efforts to develop policies and strategies that prioritize sustainability, equity, and inclusivity. It also means investing in research and development of innovative technologies that can promote sustainable development. It is essential to address the root causes of climate change and ensure that vulnerable communities are not left behind in the transition to a low-carbon economy. Achieving sustainable development that is equitable and inclusive requires a collaborative effort from all stakeholders, and urgent action is needed to address the urgent threat of climate change.

Transparency And Accountability
Guiding Principles For Climate Justice Policy

"Climate change is a global challenge that requires global solutions. we need to ensure that climate policies are transparent, accountable, and guided by the best available science. by doing so, we can build trust and ensure that our actions are effective in addressing the climate crisis."

Patricia Espinosa, Executive Secretary Of The United Nations Framework Convention On Climate Change (2018)

Climate justice demands that climate policy is guided by principles of transparency and accountability. This is because the decisions made in climate policy can significantly impact the environment, human health and well-being, and social and economic systems. Therefore, these decisions must be made in a way that is transparent and accountable to the public, and that they are based on scientific evidence and sound principles of environmental stewardship.

Transparency is essential to climate justice, as it enables stakeholders to understand the decisions being made and hold decision-makers accountable. Transparency can take many forms, such as public consultations, open data policies, and public reporting of emissions and other environmental impacts. By making information available to the public, decision-makers can build trust with stakeholders and ensure that the communities' needs and priorities guide their decisions. Main criticisms usually include the lack of enforcement (ven if transparency is mandated, there may not be sufficient enforcement mechanisms in place to ensure compliance) and data manipulation, where there is a risk that the data presented by governments and institutions may be manipulated to paint a more positive picture of their climate action than is warranted by the actual progress. However, here are three successful examples of transparency and accountability:

- The Paris Agreement: The Paris Agreement on climate change includes a transparency framework that requires all countries to report their emissions and progress towards their climate goals regularly. The agreement also includes a review process to ensure transparency and accountability. However, the effectiveness of this framework is still being debated.
- Climate Action Tracker: The Climate Action Tracker is an independent scientific analysis that tracks countries' progress towards their climate commitments. It provides a transparent and independent assessment of the effectiveness of climate policies and the credibility of climate pledges.
- Global Reporting Initiative: The Global Reporting Initiative (GRI) is an international organization that helps companies and organizations report their sustainability impacts and progress in a transparent and accountable manner. The GRI provides guidance on reporting sustainability impacts, including climate impacts.

According to a report by the World Resources Institute, only 43% of countries have mandatory national greenhouse gas inventories that are transparent and available to the public. The same report found that only 32% of countries have laws or regulations that require companies to report their emissions, and only 23% of companies in the energy and utility sector publicly disclose their emissions. A study by the University of Cambridge found that only a small percentage of large fossil fuel companies disclose their climate lobbying activities, despite the potential impact on climate policy. Accountability is also a key principle of climate justice, as it ensures that decision-makers are held responsible for their actions and that they are answerable to the public. Accountability can take many forms, such as independent audits, public hearings, and legal challenges. By holding decision-makers accountable, stakeholders can ensure that decisions are made fairly and transparently and that the impacts of those decisions are fully understood.

In addition to transparency and accountability, climate justice also demands that principles of environmental stewardship guide climate policy. This means that decision-makers should prioritize the protection of natural resources and the well-being of ecosystems and work to minimize climate change impacts. Environmental stewardship can take many forms, such as conservation programs, sustainable land use practices, and investments in renewable energy and clean technologies.

Moreover, climate justice demands that scientific evidence and sound risk management principles guide climate policy. This means that decision-makers should base their decisions on the best available scientific data and should adopt a precautionary approach to managing environmental risks. By doing so, they can ensure that decisions are made in a way that is informed by the latest scientific knowledge and that the risks associated with climate change are properly understood and addressed.

By ensuring that decisions are made transparent and accountable, and based on scientific evidence and sound principles of environmental stewardship and risk management, decision-makers can promote the well-being of communities and protect the environment for future generations. Climate justice demands that we act with urgency and determination to address the urgent challenge of climate change, and that we do so in a way that is guided by the highest standards of transparency and accountability.

Brzoska, M., Chojnacki, S., & Scheffran, J. (2019). Climate change, conflicts and cooperation in the Arctic. In The Security-Development Nexus (pp. 99-124). Springer, Cham.

Adger, W. N. (2006). Vulnerability. Global Environmental Change, 16(3), 268-281.

Stoll, R. J. (2015). Climate change and conflict in West African cities: An emerging research issue. GeoJournal, 80(4), 509-521.

United Nations Development Programme. (2017). Journey to Extremism in Africa: Drivers, Incentives and the Tipping Point for Recruitment.

Dabelko, G. D. (2009). Climate Change and Security: A Gathering Storm of Global Challenges.

Together We Can

Climate Justice As A Call For Collaborative And Inclusive Climate Action

"The climate crisis requires collaboration and partnership between governments, civil society, the private sector, and other stakeholders. climate justice provides a framework for inclusive and collaborative climate action that can help us achieve our shared goals of a more sustainable and equitable world."

Ban Ki-Moon, Former Secretary-General Of The United Nations (2017)

Climate justice promotes collaboration and partnership between different stakeholders in climate action. To address this challenge, collaboration and cooperation between various stakeholders is essential. Climate justice encourages collaboration and teamwork to ensure that all voices are heard and the needs of all stakeholders are met in climate action. Collaboration and partnership are important to climate justice because climate change is a complex problem that requires diverse expertise and perspectives to be effectively addressed. Collaboration and partnership enable different stakeholders to share knowledge, resources, and best practices, and work together towards common goals. This approach also encourages transparency, accountability, and inclusiveness, which are critical components of climate justice.

One example of collaboration and partnership in climate action is the Paris Agreement. The Paris Agreement, signed in 2015, is an international agreement that brings together 197 countries to work towards limiting global warming to well below 2 degrees

Celsius above pre-industrial levels. The agreement recognizes the importance of collaboration and partnership between different stakeholders, including governments, civil society organizations, the private sector, and the scientific community. The Paris Agreement also emphasizes the need for transparency, accountability, and inclusiveness in climate action. Another example of collaboration and partnership in climate action is the Climate Alliance of European Cities with Indigenous Rainforest Peoples. This alliance brings together European cities with indigenous communities in the Amazon rainforest to work together towards climate action. The alliance aims to promote sustainable development and protect indigenous communities and the rainforest while addressing climate change. This partnership recognizes the importance of working together across different cultures and geographic regions to address the global challenge of climate change.

Critics typically point to a perceived lack of genuine participation and representation of vulnerable communities and civil society groups in decision-making processes, leading to top-down approaches and insufficient consideration of their perspectives and needs. They may also point out potential conflicts of interest and power imbalances among stakeholders, which can undermine the effectiveness and legitimacy of collaboration efforts. In response, here are examples of this concept applied successfully:

The Global Alliance for Clean Cookstoves, which brings together governments, the private sector, NGOs and other actors to promote clean cooking solutions for households, has faced criticisms for insufficiently engaging with women and other marginalized groups most affected by indoor air pollution and lack access to clean energy. The Forest Stewardship Council, a partnership of NGOs, companies and indigenous groups to promote responsible forest management, has been criticized for allowing companies with questionable environmental and social records to participate in decision-making and certification processes, compromising the credibility and effectiveness of the partnership. The Climate and Clean Air Coalition, which is composed of governments, NGOs and private sector entities aiming to reduce short-lived climate pollutants, has been praised for promoting cooperation and innovation among diverse stakeholders, but also criticized for neglecting the voices and concerns of affected communities and failing to address the root causes of pollution.

The Climate and Clean Air Coalition (CCAC) is a voluntary partnership of governments, intergovernmental organizations, and non-governmental organizations (NGOs) that work together to reduce short-lived climate pollutants (SLCPs), including black carbon, methane, and hydrofluorocarbons (HFCs). The coalition was launched in 2012 and currently has 65 members, including 54 countries, the European Commission, and 10 international organizations. The CCAC aims to complement global efforts to reduce carbon dioxide emissions by focusing on SLCPs, which have a shorter atmospheric lifetime than carbon dioxide but are still significant contributors to climate change. The coalition works to identify and implement policies and practices that can reduce SLCP emissions across multiple sectors, including energy, transportation, agriculture, and waste management. Examples of the coalition's work include supporting the development and implementation of national SLCP reduction plans, promoting the adoption of clean cooking and heating technologies in developing countries, and promoting sustainable agriculture practices to reduce emissions of SLCPs from fertilizer and livestock. However, the CCAC has faced criticism for its focus on voluntary measures and the potential for its efforts to distract from the need to address carbon dioxide emissions. Additionally, some have raised concerns about the potential for the promotion of certain technologies, such as natural gas, to undermine efforts to transition to renewable energy sources.

- In a survey of 40 countries, only 35% of respondents thought that public engagement in climate policy was adequate, with the lowest levels of satisfaction reported in developing countries (UNDP, 2020).
- Only 4% of philanthropic funding for climate change goes to organizations based in the Global South, where

climate impacts affect vulnerable communities most (Candid, 2020).

- A study of 30 cases of multi-stakeholder partnerships found that genuine participation of civil society groups was limited in most cases, with the most marginalized and vulnerable groups often excluded (Kurian et al., 2020).
- A survey of indigenous peoples in 21 countries found that 78% of respondents did not feel that their rights were being respected in climate action (IPCC, 2019).
- A review of 48 renewable energy projects in developing countries found that less than half had involved meaningful participation of affected communities, despite the potential for renewable energy to promote social and environmental co-benefits (Woodhouse et al., 2015).

Collaboration and partnership are essential to climate justice because they promote inclusiveness, transparency, and accountability in climate action. By working together, different stakeholders can leverage their strengths and expertise to address the root causes of climate change and create a more sustainable and equitable future for all.

Leach, M., & Scoones, I. (2013). Carbon conflicts and forest landscapes in Africa. Anthropological Quarterly, 86(2), 497-520.

Ehrhart, C. (2011). Climate security and security studies: From causal ambiguity to operational discourse. Climate Change and Security, 21-36.

The World Bank. (2018). Groundswell: Preparing for Internal Climate Migration.

Think Global, Act Equitably

The Necessity Of Collective Action For Climate Justice

"Climate change is a global problem that requires global solutions. we need to work together, across borders and sectors, to address the root causes of climate change and build a more sustainable and equitable future for all."

António Guterres, Secretary-General Of The United Nations (2019)

Climate justice recognizes that climate change is a global issue that requires collective action. Addressing this challenge requires collective action from all nations and communities. Climate justice acknowledges this need for worldwide collaboration and aims to promote equity and fairness in distributing the costs and benefits of climate action. For critics, climate justice can be seen as unrealistic and overly idealistic in its call for collective action and global cooperation, especially given the current political climate and the conflicting interests of different countries and stakeholders. Some argue that climate justice overlooks the practical and economic considerations of climate policy, and may not prioritize effective solutions that can be implemented in a timely and feasible manner.

However, the Paris Agreement on climate change, signed by nearly every country in the world, is a prime example of the collective action needed to address climate change. The agreement aims to limit global temperature rise to well below 2 degrees Celsius and pursue efforts to limit it to 1.5 degrees Celsius. The Climate and Clean Air Coalition, a global partnership of governments, civil society, and private sector organizations, works to reduce short-lived climate pollutants such as black carbon and methane. The Global Climate Action Summit, held in San Francisco in 2018, brought together leaders from around the world to showcase their climate action plans and commitments.

- According to the United Nations, greenhouse gas emissions are still increasing, with the world on track to warm by 3.2 degrees Celsius by the end of the century.
- The Intergovernmental Panel on Climate Change (IPCC) has stated that limiting global warming to 1.5 degrees Celsius would require rapid, far-reaching, and unprecedented changes in all aspects of society.
- The impacts of climate change are already being felt around the world, with more frequent and intense heatwaves, droughts, floods, and storms affecting millions of people.
- Low- and middle-income countries are disproportionately affected by climate change, despite contributing the least to global greenhouse gas emissions.
- Failure to address climate change could result in significant economic losses, with the global cost of inaction estimated at $23 trillion by the end of the century.

At its core, climate justice recognizes that climate change is not simply an environmental issue, but rather a deeply interconnected issue that impacts economic, social, and political systems around the world. Climate change exacerbates existing inequalities and creates new ones, as its effects disproportionately impact vulnerable communities. This includes communities that are already facing poverty, discrimination, and marginalization, as well as those living in areas that are particularly vulnerable to the impacts of climate change, such as coastal regions and small island nations. Climate justice calls for a just transition to a low-carbon economy to address these inequalities and promote collective action. This means ensuring that the costs and benefits of this transition are fairly distributed among all communities, particularly those that are most vulnerable. It also means taking into account the historical responsibility of developed nations for greenhouse gas emissions, and providing financial and technological support to developing nations to help them transition to a low-carbon economy.

In addition to promoting a just transition, climate justice also recognizes the need for global cooperation and collaboration. Climate change is a global issue that requires action from all nations, and all countries must work together to reduce greenhouse gas emissions and mitigate the impacts of climate change. This requires a shared commitment to addressing the root causes of climate change and supporting the most vulnerable communities in adapting to its impacts. Moreover, climate justice recognizes that collective action requires a deep commitment to transparency, accountability, and cooperation. It is essential that all nations are transparent about their greenhouse gas emissions, and that they are held accountable for their commitments to reduce them. This requires cooperation and collaboration among governments, civil society, and the private sector and a commitment to open and transparent decision-making processes.

Climate justice promotes equity and fairness in the distribution of the costs and benefits of climate action, and calls for a just transition to a low-carbon economy. By recognizing the need for global cooperation and collaboration, and by promoting transparency, accountability, and cooperation, climate justice can help to ensure that the impacts of climate change are minimized, and that the transition to a sustainable future is just and equitable for all.

Buhaug, H. (2015). Climate-conflict research: some reflections on the way forward. WIREs Climate Change, 6(3), 269-275.

Brown, O., Crawford, A., & Dare, A. (2017). Climate change, migration, and conflict in Africa: A security issue.

Burke, M., Miguel, E., Satyanath, S., Dykema, J. A., & Lobell, D. B. (2009). Warming increases the risk of civil war in Africa. Proceedings of the National Academy of Sciences, 106(49), 20670-20674.

The Intergovernmental Panel on Climate Change (IPCC). (2014). Climate Change 2014: Synthesis Report.

Leaving No One Behind

Climate Justice As An Intergenerational Responsibility

"We cannot afford to ignore the impact of climate change on future generations. climate justice demands that we take responsibility for our actions and ensure that the costs of climate change are not borne by those who come after us."

Mary Robinson, Former President Of Ireland And Un Special Envoy On Climate Change (2018)

Climate justice seeks to ensure that the costs of climate change are not borne by future generations. Climate change is one of humanity's greatest challenges, with far-reaching impacts that will be felt for future generations. While it is true that some communities are already experiencing the devastating effects of climate change, it is future generations who will bear the greatest burden of this global crisis. Climate justice seeks to address this issue by ensuring that future generations do not bear the costs of climate change.

The main criticisms are generally:

Lack of Implementation: Despite the adoption of the Paris Agreement, many countries have not yet met their emissions reduction targets, and funding for climate change adaptation and mitigation efforts has been insufficient.

Economic Challenges: Another criticism is that transitioning to a low-carbon economy can pose economic challenges, particularly for developing countries. They may lack the resources to invest in renewable energy and may depend on fossil fuel exports for their economies.

In The Maldives is a low-lying island nation that is particularly vulnerable to sea-level rise caused by climate change. The country has made significant efforts to transition to renewable energy sources and has pledged to become carbon-neutral by 2030. However, they face significant challenges due to their small size and limited resources. In Indigenous communities, such as the Inuit in Canada, are also disproportionately affected by climate change. Melting permafrost and sea ice are affecting their traditional hunting and fishing practices, and they may lack the resources to adapt to these changes. For future generations they will have to deal with the consequences of our actions today. Climate justice seeks to ensure that we do not leave them with a burden that is too great to bear.

- According to the Intergovernmental Panel on Climate Change, global greenhouse gas emissions need to be reduced by 45% by 2030 in order to limit global warming to 1.5°C above pre-industrial levels.
- The impacts of climate change are already being felt around the world, with more frequent and severe heatwaves, wildfires, storms, and floods. These events are causing significant economic and human losses.
- Developing countries are particularly vulnerable to the impacts of climate change, despite contributing the least to greenhouse gas emissions. According to the World Bank, climate change could push an additional 100 million people into extreme poverty by 2030.
- Renewable energy costs are decreasing rapidly, making it increasingly competitive with fossil fuels. According to the International Renewable Energy Agency, the cost of renewable energy is now lower than the cost of new coal-fired power plants in most parts of the world.

The cost of inaction on climate change will be far greater than the cost of taking action. According to a Global Commission on the Economy and Climate report, taking action on climate change could generate $26 trillion in economic benefits by 2030, compared to the cost of inaction, which could be as high as $72 trillion by 2060.

One of the most pressing concerns of climate justice is the impact of climate change on children and young people. According to the United Nations, children are among the most vulnerable to the impacts of climate change, as they are more likely to be affected by natural disasters, water scarcity, and food insecurity. In addition, climate change is likely to have long-term impacts on children's health and well-being, with studies showing that exposure to air pollution can have negative effects on cognitive development and respiratory health. Furthermore, climate change is expected to have significant economic impacts on future generations. A report by the Intergovernmental Panel on Climate Change (IPCC) estimates that the global cost of climate change could reach $54 trillion by the end of the century if emissions are not reduced. This would have significant implications for future generations, who would inherit a world with lower economic growth and higher levels of debt. In addition to the economic costs of climate change, future generations are also at risk of losing access to natural resources essential for survival. For example, changes in weather patterns could lead to water scarcity, which would significantly impact food security and access to clean drinking water. This could have devastating consequences for communities that rely on these resources for their survival.

To address these challenges, climate justice advocates for policies and actions that ensure that future generations do not bear the costs of climate change. This includes measures to reduce greenhouse gas emissions and transition to a low-carbon economy, as well as efforts to adapt to the impacts of climate change and ensure that vulnerable communities have access to resources and support. In addition, it requires a commitment to intergenerational equity and a recognition that our actions will have profound impacts on the world we leave for future generations. This requires a commitment to intergenerational equity and a recognition of the profound impacts that our actions today will have on the world of tomorrow. As we work to address the global challenge of climate change, we must keep in mind the urgent need to protect the rights and well-being of future generations.

Center for Climate and Security. (2020). Climate Change and Security: A Gathering Storm of Global Challenges.

United Nations Security Council. (2011). Presidential Statement on Climate Change and Security.

Adger, W. N. (2010). Climate change, human well-being and insecurity. New Political Economy, 15(2), 275-292.

The Hague Declaration on Planetary Security. (2017).

Brzoska, M., Chojnacki, S., & Scheffran, J. (2019). Climate change, conflicts and cooperation in the Arctic. In The Security-Development Nexus (pp. 99-124). Springer, Cham.

Adger, W. N. (2006). Vulnerability. Global Environmental Change, 16(3), 268-281.

Stoll, R. J. (2015). Climate change and conflict in West African cities: An emerging research issue. GeoJournal, 80(4), 509-521.

Rights In The Face Of Crisis

Climate Justice And The Protection Of Human Rights

"Climate change is not just an environmental issue, it is a human rights issue. climate justice demands that we protect the rights of vulnerable communities and ensure that their voices are heard in climate action."

Michelle Bachelet, Un High Commissioner For Human Rights (2020)

Climate justice promotes the protection of human rights in the face of climate change. Climate change is an environmental problem and a social and human rights issue. As the planet continues to warm, the impacts of climate change are disproportionately felt by marginalized and vulnerable communities, who are often the least responsible for the emissions that cause it. Climate justice seeks to address these injustices by promoting the protection of human rights in the face of climate change.

Is there insufficient attention to the complexity of human rights violations? Critics argue that the human rights framework may not fully capture the complexities of climate change and its impacts on human rights. For example, climate change-induced displacement may lead to loss of livelihoods and economic opportunities, which may not be fully captured within the existing human rights framework.

Is it simply too difficult to enforcing human rights obligations? There is also a concern that enforcing human rights obligations related to climate change may be difficult due to the lack of clear legal mechanisms and the need for global cooperation.

Access to water: Climate change can impact the availability and quality of water, which is a fundamental human right. In many areas, droughts and water scarcity caused by climate change can lead to conflicts over access to water, particularly in regions where water resources are already limited.

Health: Climate change can significantly impact human health, particularly among vulnerable communities. For example, extreme heatwaves can cause heat exhaustion and heatstroke, and air pollution from burning fossil fuels can worsen respiratory illnesses.

Indigenous rights: Indigenous peoples have a unique relationship with the environment, and climate change can significantly impact their way of life and cultural heritage. Climate justice must ensure that the rights and knowledge of indigenous communities are respected and protected in climate action.

One of the most significant ways climate change threatens human rights is through its impact on the right to life. The increased frequency and intensity of extreme weather events, such as heatwaves, droughts, floods, and storms, can lead to displacement, injury, illness, and death. These events disproportionately affect those who are already marginalized, including indigenous peoples, women, children, and the elderly. Climate change also threatens the right to health, as extreme heat can exacerbate respiratory illnesses, while drought and flooding can spread waterborne diseases.

Another significant impact of climate change is its effect on food security. As temperatures rise and weather patterns become more unpredictable, agricultural yields are increasingly affected, leading to food scarcity and malnutrition. This can lead to violations of the right to food, particularly in developing countries where people rely heavily on subsistence farming.

- According to the World Health Organization, climate change is expected to cause an additional 250,000 deaths per year between 2030 and 2050 due to malnutrition, malaria, diarrhea, and heat stress.
- The United Nations estimates that by 2050, up to 1 billion people could be displaced due to climate change-induced events such as droughts, floods, and storms.
- A report by the Intergovernmental Panel on Climate Change found that indigenous peoples are disproportionately affected by climate change due to their reliance on natural resources and their vulnerability to extreme weather events.
- According to the United Nations, women are disproportionately affected by climate change due to their roles in agriculture and food security, and as primary caregivers in many communities.
- A report by the Global Justice Clinic found that climate change is likely to cause significant human rights violations, including the violation of the right to life, food, water, health, housing, and self-determination.

Climate change also has significant implications for the right to water, as droughts and water scarcity become more prevalent. This can have a particularly significant impact on indigenous peoples and other marginalized communities, who often rely on traditional water sources that are vulnerable to the impacts of climate change. In some cases, water scarcity can lead to conflict and displacement, further exacerbating the impact of climate change on human rights.

To promote climate justice and protect human rights, addressing the root causes of climate change and developing strategies to adapt to its impacts is essential. This includes reducing greenhouse gas emissions, promoting renewable energy, and investing in sustainable infrastructure. It also involves supporting vulnerable communities in adapting to the impacts of climate change, through measures such as improved water management, climate-resilient agriculture, and disaster preparedness.

Moreover, climate justice requires that human rights are at the center of climate policy decisions. This means ensuring that policies are designed to protect the most vulnerable communities and that the impacts of climate change are considered when making policy decisions. It also means that these policies should be informed by participation, accountability, and transparency principles, ensuring that marginalized communities are involved in decision-making processes.

Climate justice promotes the protection of human rights in the face of climate change. The impacts of climate change disproportionately affect marginalized and vulnerable communities, threatening their fundamental rights to life, health, food, and water. To address these challenges, it is essential to mitigate the causes of climate change and support vulnerable communities in adapting to its impacts. We can promote a just and equitable transition to a sustainable future by placing human rights at the center of climate policy decisions.

Adger, W. N. (2010). Climate change, human well-being and insecurity. New Political Economy, 15(2), 275-292.

The Hague Declaration on Planetary Security. (2017).

A Just Transition

Climate Justice And Protecting Workers And Communities

"We cannot achieve a sustainable future if we leave workers and communities behind in the transition to a low-carbon economy. climate justice demands that we ensure a just transition that protects workers and communities dependent on fossil fuels."

Sharan Burrow, General Secretary Of The International Trade Union Confederation (2019)

Climate justice ensures that the transition to a low-carbon future is just and does not harm workers or communities dependent on fossil fuels. Climate justice is an urgent and necessary goal for the global community to pursue. One of the key aspects of climate justice is the recognition that the transition to a low-carbon future must be just and equitable for all, including workers and communities that have historically been dependent on fossil fuels for their livelihoods. This complex challenge requires careful planning, investment, and collaboration across sectors and stakeholders.

One of the central concerns in transitioning to a low-carbon future is the impact on workers and communities that rely on the fossil fuel industry. The International Labour Organization estimates that the transition to a low-carbon economy could lead to the loss of 6 million jobs in the coal, oil, and gas sectors by 2030. This is a significant challenge that must be addressed just and equitably.

In order to achieve climate justice, it is important to ensure that workers and communities dependent on the fossil fuel industry are not left behind. This requires investment in job creation, training, and support for workers to transition to new, sustainable industries. It also requires investment in the infrastructure and resources necessary to support these communities in their transition.

One example of a just transition program is the Green New Deal, which aims to create jobs and invest in renewable energy and infrastructure while supporting workers and communities impacted by the fossil fuel industry. The Green New Deal includes provisions for job training, wage guarantees, and support for small businesses and local communities.

Another example of a just transition program is the Appalachian Transition Fellowship, which aims to support the transition from coal mining to sustainable industries in the Appalachian region of the United States. The fellowship provides training, mentorship, and funding for fellows to develop and implement sustainable projects in their communities. Two main criticisms of this topic are:

Lack of clear and feasible transition plans: One of the main criticisms of climate justice's focus on a just transition to a low-carbon future is that there is often a lack of clear and feasible transition plans. While it is important to ensure that workers and communities dependent on fossil fuels are not left behind in the transition to a low-carbon future, there is often a lack of clarity on how this can be achieved without harming the environment or the overall goal of reducing carbon emissions.

Limited representation and participation of affected communities: Another criticism is the limited representation and participation of affected communities, particularly those in developing countries, in decision-making processes related to climate justice. This can result in decisions that do not reflect the needs and concerns of these communities.

Just transition policies in coal mining regions: The EU's Coal Regions in Transition Initiative aims to support coal mining regions in transitioning to a low-carbon future while addressing social and economic challenges. This includes providing funding for job creation and training in new sectors and supporting community-led projects. Renewable energy projects in Indigenous communities: In Canada, the First Nations Renewable Energy Alliance is working to increase the use of renewable energy in Indigenous communities while also promoting economic development and self-determination. This includes developing community-owned renewable energy projects, providing training and education, and partnering with industry and government. Clean energy job creation in the US: The BlueGreen Alliance, a partnership between labor unions and environmental organizations, advocates for policies that support the creation of clean energy jobs while also ensuring that workers in fossil fuel industries are not left behind. This includes supporting the development of clean energy infrastructure and providing job training and transition support for workers in fossil fuel industries.

- According to the International Labour Organization, around 24 million jobs will be created globally in renewable energy and energy efficiency by 2030, more than offsetting job losses in the fossil fuel industry.
- The International Renewable Energy Agency reports that the renewable energy sector employed 11 million people worldwide in 2018, up from 10.3 million in 2017.
- A study by the National Renewable Energy Laboratory found that the renewable energy sector in the US supports more than 700,000 jobs, with solar energy accounting for the largest share of employment.
- A Global Commission on the Economy and Climate report found that transitioning to a low-carbon economy could generate $26 trillion in economic benefits by 2030.
- The International Labour Organization estimates that the cost of inaction on climate change could result in a loss of 72 million jobs globally by 2030.

Engaging with workers and communities directly impacted by the fossil fuel industry is also important to achieve climate justice and ensure a just transition. This means involving these communities in decision-making processes, listening to their concerns and priorities, and supporting their efforts to create a more sustainable and equitable future. This requires investment in job creation, training, and support for workers and communities that the fossil fuel industry has impacted. It also requires engagement with these communities and a commitment to listen to their concerns and priorities. Pursuing climate justice can create a more sustainable and equitable future for all.

Brzoska, M., Chojnacki, S., & Scheffran, J. (2019). Climate change, conflicts and cooperation in the Arctic. In The Security-Development Nexus (pp. 99-124). Springer, Cham.

Adger, W. N. (2006). Vulnerability. Global Environmental Change, 16(3), 268-281.

Stoll, R. J. (2015). Climate change and conflict in West African cities: An emerging research issue. GeoJournal, 80(4), 509-521.

United Nations Development Programme. (2017). Journey to Extremism in Africa: Drivers, Incentives and the Tipping Point for Recruitment.

Dabelko, G. D. (2009). Climate Change and Security: A Gathering Storm of Global Challenges.

Leach, M., & Scoones, I. (2013). Carbon conflicts and forest landscapes in Africa. Anthropological Quarterly, 86(2), 497-520.

Voices Of The Future

Climate Justice And Empowering Youth And Future Generations

*"We need to listen to the voices of young
people and future generations who will
inherit the consequences of our actions.
climate justice demands that we empower
youth and future generations to be part of
the solution and take action to address
climate change."*

*Mary Robinson, Former President Of Ireland
And Un Special Envoy On Climate Change
(2019)*

Climate justice seeks to ensure that the voices of youth and future generations are heard in climate action. Climate change is a complex, multi-dimensional challenge that requires urgent action on a global scale. As we work to address the causes and consequences of climate change, it is essential to ensure that the voices of youth and future generations are heard in climate action. Climate justice seeks to do just that, by promoting the participation and empowerment of young people and future generations in decision-making processes related to climate change. Some critics argue that prioritizing the voices of youth and future generations may come at the expense of the needs and concerns of present generations. Others argue that giving a prominent role to youth in climate action can create the impression that solving the climate crisis is solely their responsibility, and that they are expected to shoulder the burden of fixing a problem they did not create.

- According to the United Nations, 16-year-old climate activist Greta Thunberg has become a leading voice in the climate movement, inspiring millions of young people to demand action on climate change.
- A 2019 survey by Amnesty International found that 41% of young people aged 18-25 see climate change as the most pressing issue facing the world today.
- A 2020 Institute for Public Policy Research study found that young people in the UK are significantly more likely to prioritize climate action than older generations.
- A 2020 survey by the World Economic Forum found that young people around the world see climate change as the most pressing issue facing their generation, with 60% of respondents citing it as a top concern.
- A 2021 report by the UN Environment Programme found that the world is on track to warm by more than 3 degrees Celsius this century, putting the futures of millions of young people and future generations at risk.

The impact of climate change on the younger generation is undeniable. Young people today face an uncertain future, with the potential for more frequent and severe weather events, rising sea levels, and extreme heat waves. These events threaten their safety, security, and economic and social well-being. The current generation of young people will be the ones most affected by the consequences of climate change, and they have a right to have their voices heard in the decisions that will shape their future.

Climate justice demands that young people are included in decision-making processes related to climate action. This means ensuring that young people have a seat at the table when decisions are being made about climate policy, and that their opinions are valued and respected. It also means providing young people with access to information and resources that enable them to engage effectively in climate action, including education and training programs, mentorship opportunities, and platforms for collaboration and exchange of ideas.

Another essential aspect of climate justice is the need to ensure that future generations are considered in decision-making related to climate change. Climate change is not just a problem that we face today, but one that will continue to affect future generations. As such, it is essential to consider the long-term consequences of our actions and ensure that future generations have a say in the decisions that will affect their lives. One way to ensure that the voices of future generations are heard is through the implementation of intergenerational equity principles in climate policy. Intergenerational equity seeks to ensure that the interests of future generations are taken into account in decisions made today, recognizing that decisions made today will have consequences for future generations. By incorporating intergenerational equity into climate policy, we can ensure that the needs and concerns of future generations are not overlooked.

The global youth-led climate strikes, which began in 2018 and have continued to grow in scale and impact, are an example of youth voices being amplified in climate action. Greta Thunberg, a Swedish teenage climate activist, appointed as a UN Climate Ambassador and her prominent role in global climate negotiations is another example of youth voices being heard in climate action. The increased emphasis on intergenerational equity in the Paris Agreement, including the goal of keeping global warming below 2 degrees Celsius, is an example of the importance placed on the voices of future generations.

Young people today face an uncertain future and must have a say in the decisions that will shape their lives. By promoting the participation and empowerment of young people and future generations in decision-making processes related to climate change, we can create a more just and equitable future for all. Let us work together to create a sustainable future that recognizes the needs of both the present and future generations.

Stoll, R. J. (2015). Climate change and conflict in West African cities: An emerging research issue. GeoJournal, 80(4), 509-521.

United Nations Development Programme. (2017). Journey to Extremism in Africa: Drivers, Incentives and the Tipping Point for Recruitment.

Dabelko, G. D. (2009). Climate Change and Security: A Gathering Storm of Global Challenges.

Leach, M., & Scoones, I. (2013). Carbon conflicts and forest landscapes in Africa. Anthropological Quarterly, 86(2), 497-520.

Ehrhart, C. (2011). Climate security and security studies: From causal ambiguity to operational discourse. Climate Change and Security, 21-36.

The World Bank. (2018). Groundswell: Preparing for Internal Climate Migration.

Buhaug, H. (2015). Climate-conflict research: some reflections on the way forward. WIREs Climate Change, 6(3), 269-275.

Brown, O., Crawford, A., & Dare, A. (2017). Climate change, migration, and conflict in Africa: A security issue.

Burke, M., Miguel, E., Satyanath, S., Dykema, J. A., & Lobell, D. B. (2009). Warming increases the risk of civil war in Africa. Proceedings of the National Academy of Sciences, 106(49), 20670-20674.

The Intergovernmental Panel on Climate Change (IPCC). (2014). Climate Change 2014: Synthesis Report.

Leaving No One Behind

Climate Justice And The Transition To A Low-Carbon Future

"The transition to a low-carbon future must be just and equitable, ensuring that the most vulnerable communities are not left behind. climate justice demands that we prioritize the needs of those who are most affected by climate change and empower them to be part of the solution."

Christiana Figueres, Former Executive Secretary Of The Un Framework Convention On Climate Change (2016)

Climate justice is necessary to ensure that the most vulnerable communities are not left behind in the transition to a low-carbon future. Climate change is a global challenge that threatens the livelihoods and well-being of billions of people worldwide, especially those living in vulnerable communities. Climate justice seeks to address this challenge by ensuring that the transition to a low-carbon future is just and does not leave anyone behind. This is particularly important for the most vulnerable communities who are often the hardest hit by the impacts of climate change, such as extreme weather events, sea-level rise, and food and water insecurity.

The most vulnerable communities include low-income communities, indigenous peoples, and communities in developing countries that lack the resources and capacity to adapt to the impacts of climate change. According to the World Bank, climate change could push more than 100 million people into extreme poverty by 2030 if action is not taken. This is a grave threat to the most vulnerable communities, who often live in poverty and are already struggling to make ends meet.

Criticism 1: *Climate justice policies may increase energy prices and harm low-income communities*

One criticism of climate justice policies is that they may increase energy prices, which could disproportionately harm low-income communities. For example, a carbon tax would increase the price of fossil fuels, and low-income households typically spend a larger proportion of their income on energy than higher-income households. This could lead to increased energy poverty and job losses in industries that rely on fossil fuels.

Criticism 2: *Climate justice policies may not be implemented effectively*

Another criticism of climate justice policies is that they may not be implemented effectively. This could be due to a lack of political will or resources, corruption, or other governance challenges. For example, some countries may not have the capacity to implement climate policies effectively, which could result in the most vulnerable communities being left behind.

- 85% of the world's population lives in areas where the air quality exceeds the World Health Organization's guideline limits. Low-income communities are disproportionately affected by poor air quality, which can lead to respiratory diseases and other health issues.
- 783 million people lack access to clean water, and 2.2 billion people lack access to basic sanitation. These are primarily low-income communities, and lack of access to clean water and sanitation can lead to a range of health issues and other challenges.
- In 2019, the global energy sector accounted for 73% of greenhouse gas emissions, making it the largest contributor to climate change. Fossil fuels, such as coal, oil, and gas, are the primary source of energy in most

countries, and transitioning to renewable energy sources will be essential for reducing emissions.

- According to the International Energy Agency, renewable energy sources accounted for just 29% of global electricity generation in 2020. This highlights the need for rapid and significant investment in renewable energy to support the transition to a low-carbon future.
- According to the United Nations, climate change could push an additional 132 million people into extreme poverty by 2030. Low-income communities are already disproportionately affected by poverty and other social and economic challenges, and the impacts of climate change could exacerbate these inequalities.

Climate justice must ensure that these communities are not left behind in the transition to a low-carbon future. This means ensuring they have access to clean energy, such as solar or wind power, which can provide affordable and sustainable electricity. It also means ensuring that they have access to finance and technology to help them adapt to the impacts of climate change, such as drought-resistant crops or early warning systems for floods. In addition, climate justice seeks to ensure that the most vulnerable communities are involved in decision-making processes related to climate change. This means ensuring they have a seat at the table and their voices are heard. This is particularly important given that these communities are often the most affected by climate change but have the least influence on climate policy. Furthermore, climate justice seeks to address the root causes of climate change, such as developed countries' unsustainable consumption and production patterns. Developed countries have a historical responsibility for causing climate change, and climate justice seeks to ensure that they take responsibility for their actions and support vulnerable communities in their transition to a low-carbon future.

Hurricane Katrina

Hurricane Katrina, which devastated New Orleans in 2005, is an example of how vulnerable communities can be left behind in the face of a natural disaster. The majority of those who died or were displaced by the hurricane were low-income African American communities who lived in areas that were most vulnerable to flooding.

Flint water crisis

The Flint water crisis is another example of how vulnerable communities can be left behind in the face of environmental challenges. In 2014, the city of Flint, Michigan switched its water source to the Flint River, which was highly contaminated with lead. The majority of those affected were low-income African American communities who lacked access to clean water and had limited resources to address the crisis.

Indigenous communities in the Arctic

Indigenous communities in the Arctic are also at risk of being left behind in the transition to a low-carbon future. These communities rely on traditional subsistence activities such as hunting and fishing, which are threatened by climate change. In addition, many of these communities lack access to basic services such as running water and sanitation.

This means ensuring that they have access to clean energy, finance, and technology, and that their voices are heard in decision-making processes related to climate change. It also means addressing the root causes of climate change and ensuring that developed countries take responsibility for their actions. Only by working towards climate justice can we ensure that the transition to a low-carbon future is just and equitable for all.

United Nations. (2015). The Paris Agreement.

The Elders. (2019). Climate Change: The Security Implications for Peace and Stability.

Center for Climate and Security. (2020). Climate Change and Security: A Gathering Storm of Global Challenges.

United Nations Security Council. (2011). Presidential Statement on Climate Change and Security.

Adger, W. N. (2010). Climate change, human well-being and insecurity. New Political Economy, 15(2), 275-292.

The Hague Declaration on Planetary Security. (2017).

Brzoska, M., Chojnacki, S., & Scheffran, J. (2019). Climate change, conflicts and cooperation in the Arctic. In The Security-Development Nexus (pp. 99-124). Springer, Cham.

Adger, W. N. (2006). Vulnerability. Global Environmental Change, 16(3), 268-281.

Stoll, R. J. (2015). Climate change and conflict in West African cities: An emerging research issue. GeoJournal, 80(4), 509-521.

United Nations Development Programme. (2017). Journey to Extremism in Africa: Drivers, Incentives and the Tipping Point for Recruitment.

Dabelko, G. D. (2009). Climate Change and Security: A Gathering Storm of Global Challenges.

Leach, M., & Scoones, I. (2013). Carbon conflicts and forest landscapes in Africa. Anthropological Quarterly, 86(2), 497-520.

Ehrhart, C. (2011). Climate security and security studies: From causal ambiguity to operational discourse. Climate Change and Security, 21-36.

The World Bank. (2018). Groundswell: Preparing for Internal Climate Migration.

Buhaug, H. (2015). Climate-conflict research: some reflections on the way forward. WIREs Climate Change, 6(3), 269-275.

Brown, O., Crawford, A., & Dare, A. (2017). *Climate change, migration, and conflict in Africa: A security issue.*

Burke, M., Miguel, E., Satyanath, S., Dykema, J. A., & Lobell, D. B. (2009). *Warming increases the risk of civil war in Africa. Proceedings of the National Academy of Sciences, 106(49), 20670-20674.*

The Intergovernmental Panel on Climate Change (IPCC). (2014). *Climate Change 2014: Synthesis Report.*

Power To The People

Addressing Inequalities In Climate Decision-Making

*"Climate justice means empowering people
to be part of the solution, working together to
build a sustainable future for all."*

*Mary Robinson, Former President Of Ireland
And Un High Commissioner For Human
Rights.*

Climate justice seeks to address the inequalities in the distribution of power and influence in climate decision-making. Climate change is a global issue that requires urgent action on a scale never before seen. The effects of climate change, including rising sea levels, extreme weather patterns, and resource scarcity, threaten the social, economic, and environmental well-being of communities worldwide. As we work to address the causes and consequences of climate change, it is essential to ensure that the voices of all communities are heard in climate decision-making. Climate justice seeks to address the inequalities in the distribution of power and influence in climate decision-making to create a more just and equitable world for all.

Two main criticisms of the concept of climate justice seeking to address power and influence inequalities in climate decision-making are:

Political resistance: Addressing power and influence inequalities often involves challenging existing political structures, which may result in resistance from those who benefit from the current system. Powerful interests such as the fossil fuel industry and some governments may oppose measures to redistribute power

123

and influence in climate decision-making, leading to political obstacles and barriers to progress.

Lack of representation: Despite the growing recognition of the importance of addressing power and influence inequalities in climate decision-making, there is still a lack of representation of marginalized and vulnerable groups in decision-making spaces. This lack of representation can result in decisions that do not fully consider the needs and perspectives of these groups.

However, climate justice seeks to address the inequalities in power distribution and influence in climate decision-making. For example:

- **Indigenous knowledge and participation:** Climate justice advocates for the inclusion of indigenous peoples' knowledge and participation in decision-making on climate issues. Indigenous communities deeply understand their local ecosystems and are often disproportionately affected by climate change. Recognizing and including their knowledge and participation can lead to more effective climate policies.
- **Youth engagement:** Climate justice advocates for meaningful engagement in climate-issue decision-making. Young people will be disproportionately impacted by climate change and are often excluded from decision-making spaces. Including their voices and ideas can lead to more ambitious and effective climate policies.
- **Environmental justice:** Climate justice advocates for the inclusion of environmental justice principles in decision-making on climate issues. Environmental justice considers the disproportionate impacts of climate change on low-income and marginalized communities and seeks to address these inequalities in policy development and implementation.

Climate justice recognizes that certain communities, including indigenous peoples, low-income communities, and communities of color, are disproportionately affected by climate change. These communities often have fewer resources and less political power than others, making it difficult for them to participate fully in climate decision-making processes. Climate justice seeks to address this by promoting these communities' participation and empowerment in climate change decision-making processes. This includes providing these communities access to information and resources, ensuring their voices are heard in decision-making processes, and incorporating their perspectives into climate policies and initiatives.

In addition to addressing the inequalities in the distribution of power and influence, climate justice also seeks to address the structural and systemic factors that contribute to these inequalities. This includes recognizing and challenging the historical and ongoing injustices, such as colonization and racism, that have contributed to the marginalization of certain communities. By addressing these structural factors, climate justice seeks to create a more just and equitable world, where all communities have an equal voice in climate decision-making.

- In the United States, low-income communities and communities of color are disproportionately exposed to toxic air pollution from fossil fuel combustion, leading to higher rates of respiratory illnesses and premature deaths. (Source: Union of Concerned Scientists)
- Women in developing countries are disproportionately affected by climate change due to their roles as primary caregivers and their limited access to resources and decision-making power. (Source: UN Women)
- Indigenous peoples' traditional knowledge has contributed to the conservation and management of biodiversity in up to 80% of the world's biodiversity hotspots. (Source: UN)
- Despite young people being among the most affected by climate change, they are often excluded from climate decision-making spaces. Only 0.5% of government delegates at the 2019 UN Climate Action Summit were under the age of 30. (Source: UN Youth)

- People living in poverty contribute only a small fraction of global greenhouse gas emissions, yet are the most vulnerable to the impacts of climate change. (Source: World Bank)

One way climate justice seeks to address these inequalities is by adopting participatory decision-making processes. This approach recognizes that communities are the best experts on their own needs and challenges and seeks to incorporate their perspectives into climate decision-making processes. By including communities in the design, implementation, and evaluation of climate policies and initiatives, we can ensure that their needs and concerns are fully considered, and that policies and initiatives are tailored to meet their unique needs.

Another essential aspect of climate justice is the need to address the economic inequalities that contribute to the unequal distribution of power and influence in climate decision-making. Climate change is inextricably linked to economic and social inequality, and addressing these inequalities is essential to creating a more just and equitable world. This includes addressing income inequality, providing access to quality education and healthcare, and ensuring that communities have access to the resources and infrastructure they need to thrive.

By promoting the participation and empowerment of marginalized communities, addressing structural and systemic factors that contribute to these inequalities, and adopting participatory decision-making processes, we can create a more inclusive and equitable approach to climate decision-making. Let us work together to create a sustainable future that recognizes the needs and perspectives of all communities, and that fosters a more just and equitable world for generations to come.

Protecting Our Heritage

Climate Justice And Cultural Diversity

"Climate justice recognizes the value and importance of preserving cultural heritage and biodiversity for future generations, and the urgent need to take action to protect them from the impacts of climate change."

UNESCO, United Nations Educational, Scientific And Cultural Organization.

Climate justice is vital for the protection of cultural heritage and biodiversity. Climate change is not just a threat to the environment but also to cultural heritage and biodiversity. Climate justice recognizes the importance of protecting these resources for future generations, especially for indigenous peoples and local communities who depend on them for their cultural identity and livelihoods. Climate justice initiatives may not always fully take into account the traditional knowledge and cultural practices of Indigenous peoples, leading to unintended consequences and even harm to these communities.

Do climate justice initiatives take into account the traditional knowledge and cultural practices of Indigenous peoples, leading to unintended consequences and even harm to these communities? Does prioritizing climate justice measures may come at the expense of economic growth and job creation, particularly in industries related to fossil fuels? Not at all. Consider the Maasai community in East Africa is facing challenges due to climate change, including reduced water availability and loss of grazing land for their livestock, which are central to their way of life and cultural heritage. Again, consider that many Indigenous communities in the Amazon rainforest are facing threats to their cultural heritage and biodiversity due to deforestation, mining, and other industrial activities, often driven by international demand for resources. Or for example, coastal communities around the world are at risk of losing cultural heritage sites and traditional ways of life due to sea-level rise and other impacts of climate change.

- According to the United Nations, Indigenous peoples make up less than 5% of the world's population, but they protect around 80% of the world's biodiversity.
- A study by the World Wildlife Fund found that almost half of the world's cultural heritage sites are vulnerable to damage and destruction from climate change impacts, such as sea-level rise, extreme weather events, and wildfires.
- The Intergovernmental Panel on Climate Change (IPCC) predicts that climate change will cause significant biodiversity loss, with as many as one million species facing extinction.
- According to a report by the United Nations Environment Program, the loss of cultural heritage due to climate change is estimated to cost at least $10 billion per year globally.
- A study by the United Nations Development Programme found that Indigenous peoples are disproportionately affected by climate change, as they often depend on natural resources and live in areas particularly vulnerable to climate impacts.Cultural heritage is a community's identity, and climate change poses a significant threat to its preservation. Rising sea levels, increased temperatures, and extreme weather events are destroying cultural sites and artifacts, including ancient ruins, monuments, and historic buildings. For example, the Great Barrier Reef, a World Heritage site and one of the most biodiverse places on the planet, is under threat due to coral bleaching caused by rising sea temperatures. In 2019, the Notre-Dame Cathedral in Paris was severely damaged by a fire that was likely caused by climate change-induced drought.

Biodiversity is also essential for the planet's survival, but climate change is leading to species extinction at an unprecedented rate. The Intergovernmental Panel on Climate Change (IPCC) reports that up to one million species face extinction due to climate change, habitat loss, and other human activities. This loss of biodiversity has significant impacts on ecosystem services, such as pollination, nutrient cycling, and carbon sequestration, which are essential for human well-being and the health of the planet.

Climate justice seeks to ensure that cultural heritage and biodiversity are protected in climate action. This includes involving indigenous peoples and local communities in decision-making processes, as they have traditional knowledge and practices that can contribute to preserving these resources. Additionally, climate justice advocates for reducing greenhouse gas emissions and addressing the root causes of climate change, which are necessary to slow down the loss of biodiversity and cultural heritage.

Climate justice recognizes the interdependence between these resources, the environment, and human well-being, and seeks to ensure their protection in climate action. By working towards a low-carbon future and addressing the root causes of climate change, we can safeguard our cultural heritage and biodiversity, and ensure that they remain intact for future generations.

Center for Climate and Security. (2020). Climate Change and Security: A Gathering Storm of Global Challenges.

United Nations Security Council. (2011). Presidential Statement on Climate Change and Security.

Adger, W. N. (2010). Climate change, human well-being and insecurity. New Political Economy, 15(2), 275-292.

The Hague Declaration on Planetary Security. (2017).

Brzoska, M., Chojnacki, S., & Scheffran, J. (2019). Climate change, conflicts and cooperation in the Arctic. In The Security-Development Nexus (pp. 99-124). Springer, Cham.

Adger, W. N. (2006). Vulnerability. Global Environmental Change, 16(3), 268-281.

Stoll, R. J. (2015). Climate change and conflict in West African cities: An emerging research issue. GeoJournal, 80(4), 509-521.

United Nations Development Programme. (2017). Journey to Extremism in Africa: Drivers, Incentives and the Tipping Point for Recruitment.

Dabelko, G. D. (2009). Climate Change and Security: A Gathering Storm of Global Challenges.

Leach, M., & Scoones, I. (2013). Carbon conflicts and forest landscapes in Africa. Anthropological Quarterly, 86(2), 497-520.

Ehrhart, C. (2011). Climate security and security studies: From causal ambiguity to operational discourse. Climate Change and Security, 21-36.

The World Bank. (2018). Groundswell: Preparing for Internal Climate Migration.

Buhaug, H. (2015). Climate-conflict research: some reflections on the way forward. WIREs Climate Change, 6(3), 269-275.

Brown, O., Crawford, A., & Dare, A. (2017). Climate change, migration, and conflict in Africa: A security issue.

Burke, M., Miguel, E., Satyanath, S., Dykema, J. A., & Lobell, D. B. (2009). Warming increases the risk of civil war in Africa. Proceedings of the National Academy of Sciences, 106(49), 20670-20674.

The Intergovernmental Panel on Climate Change (IPCC). (2014). Climate Change 2014: Synthesis Report.

Power To The People

Promoting Equitable Access To Renewable Energy

"Climate change is a global problem with grave implications: environmental, social, economic, political and for the distribution of goods. it represents one of the principal challenges facing humanity in our day."

Pope Francis, Laudato Si' (2015)

Climate justice promotes the use of renewable energy that is accessible and affordable for all. Climate justice is a critical aspect of the fight against climate change, and one key component of this is the promotion of renewable energy that is accessible and affordable for all. This is because transitioning to renewable energy is essential for reducing greenhouse gas emissions and mitigating the impacts of climate change. However, it is equally important to ensure that this transition is equitable and fair, particularly for those who may be most vulnerable to the impacts of climate change and who may have been historically marginalized in the energy sector.

Can the high upfront costs of renewable energy systems hinder low-income communities from accessing clean energy? Critics say that while renewable energy systems may be cheaper in the long run, the initial costs can be prohibitive for those with limited financial resources. What about the production of renewable energy infrastructure such as wind turbines and solar panels, can these have negative environmental and social impacts, particularly in areas where they are manufactured or installed? The Indian government's "Saubhagya" scheme aims to provide electricity to all households in the country through grid extension and renewable energy solutions. However, the upfront costs of solar home systems remain a barrier for many low-income households. In the United States, low-income households are often disproportionately affected by energy insecurity and high energy costs. The expansion of community solar programs can provide access to clean energy without the upfront costs of installing solar panels on individual homes. The African Development Bank has set a goal of providing universal access to electricity in Africa by 2025. To achieve this, they are prioritizing the development of renewable energy infrastructure that is affordable and accessible to all, including low-income communities.

- According to the International Energy Agency, over 770 million people globally still lack access to electricity, with a large proportion residing in low-income countries.
- In the United States, low-income households spend a larger percentage of their income on energy costs compared to higher-income households, making energy bills a significant financial burden for many.
- The cost of solar panels has decreased by 90% over the last decade, making it more affordable for households and communities to transition to renewable energy.
- According to the World Bank, investing in renewable energy can create new jobs and stimulate economic growth. For example, the renewable energy sector in Kenya has the potential to create over 200,000 jobs by 2030.
- A study by the National Renewable Energy Laboratory found that community solar projects can provide up to 50% energy cost savings for low-income households, providing a significant financial benefit for those who may otherwise struggle to access clean energy.

One of the key benefits of renewable energy is that it is typically much cleaner than traditional fossil fuel-based energy sources, and it produces significantly fewer greenhouse gas emissions. This is particularly important in the context of climate change, as reducing emissions is essential for mitigating the most catastrophic impacts of a changing climate. Renewable energy can also be much more sustainable than fossil fuels, which are a finite resource that will eventually run out.

Another important aspect of promoting renewable energy is ensuring that it is accessible and affordable for all. This is particularly important for those who may be most vulnerable to the impacts of climate change, such as low-income communities and communities of color. These communities have historically been marginalized in the energy sector and have often been subject to higher energy costs and lower access to clean energy resources.

To address these inequalities, climate justice advocates for policies and programs prioritizing these communities' needs and ensuring they have access to clean, affordable energy. This might include programs that provide financial support for the installation of renewable energy systems, or policies that prioritize the development of renewable energy resources in underserved communities. In addition to being accessible and affordable, renewable energy must also be reliable and resilient in the face of climate impacts. This means ensuring that renewable energy infrastructure is designed to withstand extreme weather events and other climate impacts. It also means investing in the development of new renewable energy technologies that can provide reliable power even in challenging environmental conditions.

By transitioning away from fossil fuels and towards clean, sustainable energy sources, we can reduce greenhouse gas emissions, mitigate the impacts of climate change, and promote greater equity and fairness in the energy sector. To achieve this goal, we must prioritize marginalized communities' needs and ensure they have access to the resources and support they need to transition to a clean energy future.

United Nations Development Programme. (2017). Journey to Extremism in Africa: Drivers, Incentives and the Tipping Point for Recruitment.

Dabelko, G. D. (2009). *Climate Change and Security: A Gathering Storm of Global Challenges.*

Leach, M., & Scoones, I. (2013). *Carbon conflicts and forest landscapes in Africa. Anthropological Quarterly, 86(2), 497-520.*

Ehrhart, C. (2011). *Climate security and security studies: From causal ambiguity to operational discourse. Climate Change and Security, 21-36.*

The World Bank. (2018). *Groundswell: Preparing for Internal Climate Migration.*

Buhaug, H. (2015). *Climate-conflict research: some reflections on the way forward. WIREs Climate Change, 6(3), 269-275.*

Brown, O., Crawford, A., & Dare, A. (2017). *Climate change, migration, and conflict in Africa: A security issue.*

Equity In Action

The Importance Of Climate Justice For An Effective Global Response

"Climate justice is a necessary condition for the effectiveness of climate action. without justice, the global response to climate change will remain fragmented and inadequate, failing to address the root causes of the problem and leaving vulnerable communities behind."

(Hassan, 2019)

Climate justice is necessary to ensure that the global response to climate change is effective and equitable. Climate change is one of our most significant global challenges, and it requires urgent action from all nations and peoples. However, the impacts of climate change are not equally distributed, and vulnerable communities, especially those in low-income countries, are often the most affected. Climate justice seeks to ensure that the global response to climate change is effective and equitable, with a focus on addressing the disproportionate impacts of climate change on vulnerable populations.

Two main criticisms of climate justice are:

1. *Lack of clarity and agreement on what constitutes climate justice*: There is no clear and agreed-upon definition of climate justice, which can lead to confusion and disagreements about what specific policies or actions are required to achieve it.
2. *Challenges in implementing climate justice policies*: Climate justice policies often require significant changes

to current economic and political systems, which can be difficult to achieve. These changes can face resistance from powerful interests, which can make it challenging to implement effective climate justice policies.

The disproportionate impact of climate change on vulnerable communities: Low-income communities, people of color, and Indigenous communities are often disproportionately impacted by climate change, despite contributing the least to the problem. This can include increased exposure to extreme weather events, water scarcity, and food insecurity. The unequal distribution of the costs and benefits of climate action: The costs of transitioning to a low-carbon economy are often borne by those who can least afford it, while wealthier communities often feel the benefits. For example, low-income households may face higher energy bills due to renewable energy subsidies, while wealthier households may benefit from increased property values due to the installation of solar panels. The need for developed countries to take responsibility for historical emissions: Developed countries have historically been responsible for the majority of global greenhouse gas emissions, but many developing countries are now experiencing the impacts of climate change. Climate justice advocates argue that developed countries have a moral obligation to take action to address climate change and support vulnerable communities around the world.

- According to the United Nations, people in developing countries are four times more likely to be affected by climate-related disasters than people in developed countries.
- A report by Oxfam found that the world's richest 10% of people are responsible for more than half of global greenhouse gas emissions, while the poorest 50% are responsible for just 10%.

- Indigenous peoples, who make up 5% of the world's population, protect 80% of global biodiversity, including many carbon-rich forests. However, their lands and rights are often threatened by extractive industries, which contribute to climate change.
- The International Energy Agency estimates that global subsidies for fossil fuels reached $5.9 trillion in 2020, with the majority of these subsidies going to wealthy countries.
- A study published in the journal Nature found that climate change could lead to a 10% decrease in global GDP by the end of the century, with the largest impacts felt in the poorest countries.

One of the key principles of climate justice is the recognition of historical responsibility. Developed countries, with their long history of greenhouse gas emissions, have a greater responsibility for addressing climate change. According to the Intergovernmental Panel on Climate Change (IPCC), developed countries accounted for over 70% of global carbon dioxide emissions between 1850 and 2010. This historical responsibility must be recognized and addressed through climate action, including financial and technological support for vulnerable communities. Another important aspect of climate justice is the need for global collaboration and cooperation. Climate change is a global issue, and no country can address it alone. Effective global response to climate change requires cooperation and partnership between all countries, with a focus on the needs and priorities of the most vulnerable communities. This collaboration must also include the sharing of knowledge and expertise, especially in the areas of clean energy and sustainable development.

Climate justice also recognizes the interconnection between social justice and environmental sustainability. The impacts of climate change are not only environmental but also social and economic, with the potential to exacerbate existing inequalities and deepen poverty. Climate justice seeks to ensure that vulnerable communities are not left behind in the transition to a low-carbon future, and that the costs of climate change are not borne by the most vulnerable. Furthermore, climate justice is essential for the protection of cultural heritage and biodiversity. Climate change poses a significant threat to the preservation of cultural heritage sites, such as historic landmarks and monuments, and natural ecosystems. These resources are essential for the livelihoods and well-being of vulnerable communities, and their protection is necessary to ensure a just transition to a low-carbon future.

This requires the recognition of historical responsibility, global collaboration and cooperation, the interconnection between social justice and environmental sustainability, and the protection of cultural heritage and biodiversity. Through climate justice, we can build a sustainable and just future for all people and the planet.

Concluding Call

The Climate Crisis: A Global Challenge

Why Urgent Action Is Needed to Protect Our Planet and Future

"We face a global climate emergency. The climate crisis is deepening. We are already seeing the devastating impacts of climate change, from wildfires to droughts to hurricanes and typhoons. This is not a problem that can be solved by one country alone. It requires urgent action by all nations, working together to reduce greenhouse gas emissions, protect vulnerable communities, and build a sustainable future for all."

United Nations Secretary-General António Guterres

Climate change is a global challenge that requires our urgent action. Climate justice is a framework that recognizes the disproportionate impacts of climate change on vulnerable communities and seeks to address the social and economic inequalities that contribute to climate change. Climate justice demands that the transition to a low-carbon future is just and does not harm workers or communities dependent on fossil fuels. It also promotes collaboration and partnership between different stakeholders in climate action, and recognizes the interconnection between social justice and environmental sustainability. Climate justice is necessary to ensure that the global response to climate change is effective and equitable, and that the costs and benefits of climate action are distributed fairly. By embracing the principles of climate justice, we can work towards a sustainable and just future for all.

United Nations. (2015). The Paris Agreement.

The Elders. (2019). Climate Change: The Security Implications for Peace and Stability.

Center for Climate and Security. (2020). Climate Change and Security: A Gathering Storm of Global Challenges.

United Nations Security Council. (2011). Presidential Statement on Climate Change and Security.

Adger, W. N. (2010). Climate change, human well-being and insecurity. New Political Economy, 15(2), 275-292.

The Hague Declaration on Planetary Security. (2017).

Brzoska, M., Chojnacki, S., & Scheffran, J. (2019). Climate change, conflicts and cooperation in the Arctic. In The Security-Development Nexus (pp. 99-124). Springer, Cham.

Adger, W. N. (2006). Vulnerability. Global Environmental Change, 16(3), 268-281.

Stoll, R. J. (2015). Climate change and conflict in West African cities: An emerging research issue. GeoJournal, 80(4), 509-521.

United Nations Development Programme. (2017). Journey to Extremism in Africa: Drivers, Incentives and the Tipping Point for Recruitment.

Dabelko, G. D. (2009). Climate Change and Security: A Gathering Storm of Global Challenges.

Leach, M., & Scoones, I. (2013). Carbon conflicts and forest landscapes in Africa. Anthropological Quarterly, 86(2), 497-520.

Ehrhart, C. (2011). *Climate security and security studies: From causal ambiguity to operational discourse. Climate Change and Security, 21-36.*

The World Bank. (2018). *Groundswell: Preparing for Internal Climate Migration.*

Buhaug, H. (2015). *Climate-conflict research: some reflections on the way forward. WIREs Climate Change, 6(3), 269-275.*

Brown, O., Crawford, A., & Dare, A. (2017). *Climate change, migration, and conflict in Africa: A security issue.*

Burke, M., Miguel, E., Satyanath, S., Dykema, J. A., & Lobell, D. B. (2009). *Warming increases the risk of civil war in Africa. Proceedings of the National Academy of Sciences, 106(49), 20670-20674.*

The Intergovernmental Panel on Climate Change (IPCC). (2014). *Climate Change 2014: Synthesis Report.*

United Nations. (2015). *The Paris Agreement.*

The Elders. (2019). *Climate Change: The Security Implications for Peace and Stability.*

Center for Climate and Security. (2020). *Climate Change and Security: A Gathering Storm of Global Challenges.*

United Nations Security Council. (2011). *Presidential Statement on Climate Change and Security.*

Adger, W. N. (2010). *Climate change, human well-being and insecurity. New Political Economy, 15(2), 275-292.*

The Hague Declaration on Planetary Security. (2017).

Brzoska, M., Chojnacki, S., & Scheffran, J. (2019). *Climate change, conflicts and cooperation in the Arctic. In The Security-Development Nexus (pp. 99-124). Springer, Cham.*

Adger, W. N. (2006). *Vulnerability. Global Environmental Change, 16(3), 268-281.*

Stoll, R. J. (2015). *Climate change and conflict in West African cities: An emerging research issue. GeoJournal, 80(4), 509-521.*

United Nations Development Programme. (2017). *Journey to Extremism in Africa: Drivers, Incentives and the Tipping Point for Recruitment.*

Dabelko, G. D. (2009). *Climate Change and Security: A Gathering Storm of Global Challenges.*

Leach, M., & Scoones, I. (2013). *Carbon conflicts and forest landscapes in Africa. Anthropological Quarterly, 86(2), 497-520.*

Ehrhart, C. (2011). *Climate security and security studies: From causal ambiguity to operational discourse. Climate Change and Security, 21-36.*

The World Bank. (2018). *Groundswell: Preparing for Internal Climate Migration.*

Buhaug, H. (2015). *Climate-conflict research: some reflections on the way forward. WIREs Climate Change, 6(3), 269-275.*

Brown, O., Crawford, A., & Dare, A. (2017). *Climate change, migration, and conflict in Africa: A security issue.*

Burke, M., Miguel, E., Satyanath, S., Dykema, J. A., & Lobell, D. B. (2009). *Warming increases the risk of civil war in Africa. Proceedings of the National Academy of Sciences, 106(49), 20670-20674.*

The Intergovernmental Panel on Climate Change (IPCC). (2014). *Climate Change 2014: Synthesis Report.*

United Nations. (2015). *The Paris Agreement.*

The Elders. (2019). *Climate Change: The Security Implications for Peace and Stability.*

Center for Climate and Security. (2020). *Climate Change and Security: A Gathering Storm of Global Challenges.*

United Nations Security Council. (2011). *Presidential Statement on Climate Change and Security.*

Adger, W. N. (2010). *Climate change, human well-being and insecurity. New Political Economy, 15(2), 275-292.*

The Hague Declaration on Planetary Security. (2017).

Brzoska, M., Chojnacki, S., & Scheffran, J. (2019). *Climate change, conflicts and cooperation in the Arctic. In The Security-Development Nexus (pp. 99-124). Springer, Cham.*

Adger, W. N. (2006). *Vulnerability. Global Environmental Change, 16(3), 268-281.*

Stoll, R. J. (2015). Climate change and conflict in West African cities: An emerging research issue. GeoJournal, 80(4), 509-521.

United Nations Development Programme. (2017). Journey to Extremism in Africa: Drivers, Incentives and the Tipping Point for Recruitment.

Dabelko, G. D. (2009). Climate Change and Security: A Gathering Storm of Global Challenges.

Leach, M., & Scoones, I. (2013). Carbon conflicts and forest landscapes in Africa. Anthropological Quarterly, 86(2), 497-520.

Ehrhart, C. (2011). Climate security and security studies: From causal ambiguity to operational discourse. Climate Change and Security, 21-36.

The World Bank. (2018). Groundswell: Preparing for Internal Climate Migration.

Buhaug, H. (2015). Climate-conflict research: some reflections on the way forward. WIREs Climate Change, 6(3), 269-275.

Brown, O., Crawford, A., & Dare, A. (2017). Climate change, migration, and conflict in Africa: A security issue.

Burke, M., Miguel, E., Satyanath, S., Dykema, J. A., & Lobell, D. B. (2009). Warming increases the risk of civil war in Africa. Proceedings of the National Academy of Sciences, 106(49), 20670-20674.

The Intergovernmental Panel on Climate Change (IPCC). (2014). Climate Change 2014: Synthesis Report.

United Nations. (2015). The Paris Agreement.

The Elders. (2019). Climate Change: The Security Implications for Peace and Stability.

Center for Climate and Security. (2020). Climate Change and Security: A Gathering Storm of Global Challenges.

United Nations Security Council. (2011). Presidential Statement on Climate Change and Security.

Adger, W. N. (2010). Climate change, human well-being and insecurity. New Political Economy, 15(2), 275-292.

The Hague Declaration on Planetary Security. (2017).

Brzoska, M., Chojnacki, S., & Scheffran, J. (2019). *Climate change, conflicts and cooperation in the Arctic. In The Security-Development Nexus (pp. 99-124). Springer, Cham.*

Adger, W. N. (2006). *Vulnerability. Global Environmental Change, 16(3), 268-281.*

Stoll, R. J. (2015). *Climate change and conflict in West African cities: An emerging research issue. GeoJournal, 80(4), 509-521.*

United Nations Development Programme. (2017). *Journey to Extremism in Africa: Drivers, Incentives and the Tipping Point for Recruitment.*

Dabelko, G. D. (2009). *Climate Change and Security: A Gathering Storm of Global Challenges.*

Leach, M., & Scoones, I. (2013). *Carbon conflicts and forest landscapes in Africa. Anthropological Quarterly, 86(2), 497-520.*

Ehrhart, C. (2011). *Climate security and security studies: From causal ambiguity to operational discourse. Climate Change and Security, 21-36.*

The World Bank. (2018). *Groundswell: Preparing for Internal Climate Migration.*

Buhaug, H. (2015). *Climate-conflict research: some reflections on the way forward. WIREs Climate Change, 6(3), 269-275.*

Brown, O., Crawford, A., & Dare, A. (2017). *Climate change, migration, and conflict in Africa: A security issue.*

Burke, M., Miguel, E., Satyanath, S., Dykema, J. A., & Lobell, D. B. (2009). *Warming increases the risk of civil war in Africa. Proceedings of the National Academy of Sciences, 106(49), 20670-20674.*

The Intergovernmental Panel on Climate Change (IPCC). (2014). *Climate Change 2014: Synthesis Report.*

United Nations. (2015). *The Paris Agreement.*

The Elders. (2019). *Climate Change: The Security Implications for Peace and Stability.*

Center for Climate and Security. (2020). *Climate Change and Security: A Gathering Storm of Global Challenges.*

United Nations Security Council. (2011). *Presidential Statement on Climate Change and Security.*

Adger, W. N. (2010). Climate change, human well-being and insecurity. New Political Economy, 15(2), 275-292.

The Hague Declaration on Planetary Security. (2017).

Brzoska, M., Chojnacki, S., & Scheffran, J. (2019). Climate change, conflicts and cooperation in the Arctic. In The Security-Development Nexus (pp. 99-124). Springer, Cham.

Adger, W. N. (2006). Vulnerability. Global Environmental Change, 16(3), 268-281.

Stoll, R. J. (2015). Climate change and conflict in West African cities: An emerging research issue. GeoJournal, 80(4), 509-521.

United Nations Development Programme. (2017). Journey to Extremism in Africa: Drivers, Incentives and the Tipping Point for Recruitment.

Dabelko, G. D. (2009). Climate Change and Security: A Gathering Storm of Global Challenges.

Leach, M., & Scoones, I. (2013). Carbon conflicts and forest landscapes in Africa. Anthropological Quarterly, 86(2), 497-520.

Ehrhart, C. (2011). Climate security and security studies: From causal ambiguity to operational discourse. Climate Change and Security, 21-36.

The World Bank. (2018). Groundswell: Preparing for Internal Climate Migration.

Buhaug, H. (2015). Climate-conflict research: some reflections on the way forward. WIREs Climate Change, 6(3), 269-275.

Brown, O., Crawford, A., & Dare, A. (2017). Climate change, migration, and conflict in Africa: A security issue.

www.ingramcontent.com/pod-product-compliance
Lightning Source LLC
Chambersburg PA
CBHW070552220526
45467CB00003B/1176